WITTGENSTEIN'S LECTURES

Titles on Metaphysics and Epistemology in Prometheus's Great Books in Philosophy Series

ARISTOTLE
- *De Anima*
- *The Metaphysics*

FRANCIS BACON
- *Essays*

GEORGE BERKELEY
- *Three Dialogues Between Hylas and Philonous*

W. K. CLIFFORD
- *The Ethics of Belief and Other Essays*

RENÉ DESCARTES
- *Discourse on Method* and *The Meditations*

JOHN DEWEY
- *How We Think*
- *The Influence of Darwin on Philosophy and Other Essays*

EPICURUS
- *The Essential Epicurus: Letters, Principal Doctrines, Vatican Sayings, and Fragments*

SIDNEY HOOK
- *The Quest for Being*

DAVID HUME
- *An Enquiry Concerning Human Understanding*
- *Treatise of Human Nature*

WILLIAM JAMES
- *The Meaning of Truth*
- *Pragmatism*

IMMANUEL KANT
- *The Critique of Judgment*
- *Critique of Practical Reason*
- *Critique of Pure Reason*

GOTTFRIED WILHELM LEIBNIZ
- *Discourse on Metaphysics* and *The Monadology*

JOHN LOCKE
- *An Essay Concerning Human Understanding*

CHARLES S. PEIRCE
- *The Essential Writings*

PLATO
- *The Euthyphro, Apology, Crito,* and *Phaedo*
- *Lysis, Phaedrus,* and *Symposium: Plato on Homosexuality*

BERTRAND RUSSELL
- *The Problems of Philosophy*

GEORGE SANTAYANA
- *The Life of Reason*

SEXTUS EMPIRICUS
- *Outlines of Pyrrhonism*

LUDWIG WITTGENSTEIN
- *Wittgenstein's Lectures: Cambridge, 1932–1935*

See the back of this volume for a complete list of titles in Prometheus's Great Books in Philosophy and Great Minds series.

WITTGENSTEIN'S LECTURES

Cambridge, 1932-1935

From the Notes of
Alice Ambrose and Margaret Macdonald

Edited by

ALICE AMBROSE

GREAT BOOKS IN PHILOSOPHY

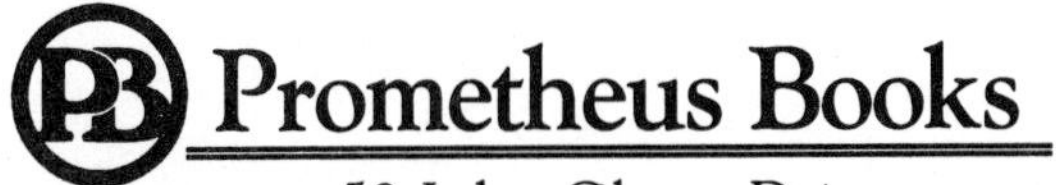

59 John Glenn Drive
Amherst, New York 14228-2197

Published 2001 by Prometheus Books

Inquiries should be addressed to
Prometheus Books
59 John Glenn Drive
Amherst, New York 14228–2197
VOICE: 716–691–0133, ext. 207
FAX: 716–564–2711
WWW.PROMETHEUSBOOKS.COM

05 04 03 02 01   5 4 3 2 1

First U.S. paperback edition, Phoenix Books Division of
The University of Chicago Press (1982).

Library of Congress Cataloging-in-Publication Data

Wittgenstein, Ludwig, 1889–1951.
[Lectures, Cambridge, 1932–1935]
Wittgenstein's lectures, Cambridge, 1932–1935 : from the notes of Alice Ambrose and Margaret Macdonald / edited by Alice Ambrose.
p. cm. — (Great books in philosophy)
Originally published: Totowa, N.J. : Rowman and Littlefield, 1979.
Includes bibliographical references.
ISBN 1–57392–875–5
1. Philosophy. 2. Mathematics—Philosophy. I. Ambrose, Alice, 1906–2001. II. Macdonald, Margaret. III. Title. IV. Series.

B3376 .W561 2001
192—dc21   2001019037

Printed in the United States of America on acid-free paper

LUDWIG WITTGENSTEIN was born in Vienna, Austria, on April 26, 1889, the youngest of eight children of a leading Austrian steelmaker. Educated at home until age fourteen, Wittgenstein then studied mathematics and natural sciences in an Austrian school, and later studied mechanical engineering for two years in Berlin, Germany. In 1908 he engaged in aeronautical research in England. He registered as a research student at the University of Manchester, where he worked in an engineering laboratory. Bertrand Russell's book *The Principles of Mathematics* (1903) had a profound influence on Wittgenstein, and he left the university in 1911 to study mathematical logic with Russell at Cambridge.

Wittgenstein remained at Cambridge until 1913, when he traveled to Skjolden, Norway, where he secluded himself to continue his study of logic. Upon the outbreak of World War I, he enlisted in the Austrian army, eventually serving as an artillery officer on both the eastern and western fronts. Throughout the war, Wittgenstein continued to work on the problems of philosophy and logic, recording his thoughts in notebooks. At the end of the war, he sent his manuscript to Russell, who found a publisher for his work.

Published in 1921, the *Tractatus Logico-Philosophicus* has been universally admired. This work covers a vast range of topics: the nature of language; the limits of what can be said; logic, ethics, and philosophy; causality and induction; the self and the will; death and the mystical; good and evil.

Upon returning to civilian life in 1919, Wittgenstein gave away the large fortune he had inherited from his father, and lived a frugal and simple life. Feeling that he could contribute nothing more to philosophy after publication of the *Tractatus*, he searched for a new vocation, first teaching elementary school in Austria for several years, then turning to gardening and architecture.

In 1929 Wittgenstein felt that once again he could do creative work in philosophy. He returned to Cambridge in 1929, where he was made a fellow of Trinity College. Through his lectures and the wide circulation of notes taken by his students, he gradually came to exert a powerful influence on philosophical thought throughout the English-speaking world. In 1936 he began his second major work, *Philosophical Investigations*.

In 1939 Wittgenstein was appointed to the chair in philosophy at

Cambridge University. During World War II he left Cambridge and worked as a porter in Guy's hospital in London and then as a laboratory assistant in the Royal Victoria Infirmary. In 1944 he returned to Cambridge as professor of philosophy, but resigned his chair in 1947. He completed the *Investigations* in 1949, which he instructed should be published only after his death.

Frequently ill during his remaining years, Wittgenstein was diagnosed with cancer in 1949. He died in Cambridge, England, on April 29, 1951. *Philosophical Investigations* was published in 1953.

Contents

For Morris

Editor's Preface

The difficulties I have encountered in editing these notes, which were taken from 1932 to 1935 when I attended Wittgenstein's lectures, have been multiplied by my not having notes of other members of the same classes against which to check my own, except for the year 1934–35, when Dr. Margaret Macdonald and I shared our notes. The first draft I made on leaving Cambridge was a compilation of her notes and mine for that year, and the original draft of her notes was kindly made available to me by Mr. Rush Rhees, together with scattered notes which Wittgenstein made at the time he gave the lectures of the Easter term of 1934–35. For the lectures of 1932–33, my first year in Cambridge, I have had to depend solely on my notes and my memory, both of lectures of that year entitled "Philosophy for Mathematicians" and of lectures to a larger class throughout the year.

The so-called Yellow Book consists of notes taken during the year 1933–34 by Ms. Margaret Masterman and myself of lectures and informal discussions during intervals in the dictation of *The Blue Book*. What appears here as Part II includes only my own notes, not those of Ms. Masterman. Notes taken on the same material by Francis Skinner, now deceased, were included in the Yellow Book but formed only a small part of the total. It seemed to me better on the whole to use notes for which I alone was responsible.

My concern has been to present a connected and faithful account of what Wittgenstein said. Some of the people who might have aided me in this endeavor by providing me with notes for comparison with my own are now dead. The notes of others who were in the classes were too meagre to be used. Wittgenstein's published writings, wherever they contain material overlapping with my own, are of course confirmation. Where materials are not treated in the manuscripts now published, I have simply recorded what I had, in order to make related texts available to philosophers. A few of the notes were too confused or obscure to permit of intelligible reconstruction, and these I have omitted, my reason being that my notes rather than Wittgenstein's lectures were at fault. I have also omitted notes which were duplications of well-known material already published in his *Tractatus*. I think he would have been in agreement, and that his own deletions would have been far more extensive.

It is reasonable to suppose that the lectures entitled "Philosophy for Mathematicians" followed fairly closely the materials published in *Philosophische Bemerkungen* and in *Philosophische Grammatik*. For this set of lectures was given in 1932–33, at about the time when Wittgenstein had written, or was writing, these two books. The notes of these lectures are placed after those of the Easter term of 1934–35 rather than in their chronological order because of their connection with the subject matter of that term. I am much indebted to two mathematicians, Dr. G. T. Kneebone of Bedford College, London, and Professor H. S. M. Coxeter of Toronto University, who read certain mathematical parts of these notes.

As might be expected, problems treated in the Yellow Book are for the most part those treated in *The Blue Book*. Their main value lies in their sometimes being better stated than in *The Blue Book* dictation, though certain things which I have included and which I think are important are not to be found elsewhere. In addition to taking notes of lectures of 1933–34, Ms. Masterman and I took notes of his informal discussions in the intervals between dictation when, as he thought, and sometimes regretted, no record had been made of what he said. Subsequently, explicit permission was given us to continue with note-taking of his informal discussions.

I have made little attempt here to collate what I have with his published works on a given topic, though in a few cases I have made references. Nor have I attempted always to follow the exact order of his presentation in lectures. Those who know his style of lecturing will remember that a topic often recurred, if only in a recapitulation, in a subsequent lecture, and that even within a lecture comments on some matter whose relevance was not clear to his class would be noted, dropped, and sometimes taken up again later. I had at first thought to delete some of the repetitions which occurred throughout a term or occurred in another year in lectures to a different class, but for the most part I have not done this. In some cases I have brought together widely separated remarks not integral to the discussion at the time, for example, on existence proofs and formalism. Throughout, I have indicated when relevant material from other notes of mine has been incorporated in the treatment of a given problem. In a number of places I have inserted comments of my own, indicated by square brackets, when connections within the lecture material seemed lacking. The materials of the first two years are usually given here in the order in which he took them up. The numbering indicates their order and is intended for convenient reference. Divisions in the notes of the last year coincide with the successive lectures in each of the academic terms.

What remains after culling and revising is, I think, substantially correct. Notes taken at the beginning of 1932–33, during my first weeks in Cambridge, both of the lectures called "Philosophy for Mathematicians", and of lectures called "Philosophy" which were given to a different class, are the least satisfactory. Only the fact that I wrote out most of the notes in full sentences shortly after leaving Cambridge saves them from the inaccuracy from which they might have suffered these many years later. In the final assembling and editing of the notes of the three years I was fortunate to have the help of Professor Morris Lazerowitz, who gave over to this work three months of his own time; and I am deeply indebted to him for what he has contributed to the readability and clarity of the resultant draft.

Editing these lectures has been rewarding for me in a special and personal way. It has enabled me to return in memory to the hours I spent in lectures and dictation in Wittgenstein's rooms in Whewell's Court, and to the hours of discussion I had with G. E. Moore in his study at 86 Chesterton Road, often on topics Wittgenstein was discussing. It has made it possible for me to recapture some of the intellectual excitement which permeated the atmosphere in those years.

Conway, Massachusetts

PART I

Philosophy

Wittgenstein's Lectures

1932–33

From the notes of Alice Ambrose

Philosophy

1932–33

1 I am going to exclude from our discussion questions which are answered by experience. Philosophical problems are not solved by experience, for what we talk about in philosophy are not facts but things for which facts are useful. Philosophical trouble arises through seeing a system of rules and seeing that things do not fit it. It is like advancing and retreating from a tree stump and seeing different things. We go nearer, remember the rules, and feel satisfied, then retreat and feel dissatisfied.

2 Words and chess pieces are analogous; knowing how to use a word is like knowing how to move a chess piece. Now how do the rules enter into playing the game? What is the difference between playing the game and aimlessly moving the pieces? I do not deny there is a difference, but I want to say that knowing how a piece is to be used is not a particular state of mind which goes on while the game goes on. The meaning of a word is to be defined by the rules for its use, not by the feeling that attaches to the words.

"How is the word used?" and "What is the grammar of the word?" I shall take as being the same question.

The phrase, "bearer of the word", [standing for what one points to in giving an ostensive definition], and "meaning of the word" have entirely different grammars; the two are not synonymous. To explain a word such as "red" by pointing to something gives but one rule for its use, and in cases where one cannot point, rules of a different sort are given. All the rules together give the meaning, and these are not fixed by giving an ostensive definition. The rules of grammar are entirely independent of one another. Two words have the same meaning if they have the same rules for their use.

Are the rules, for example, $\sim\sim p = p$ for negation, responsible to the meaning of a word? No. The rules constitute the meaning, and are not responsible to it. The meaning changes when one of its rules changes. If, for example, the game of chess is defined in terms of its rules, one cannot say the game changes if a rule for moving a piece were changed. Only when we are speaking of the history of the game can we talk of change. Rules are arbitrary in the sense that they are not responsible to some sort of reality—they are not similar to natural laws; nor are they responsible to some meaning the word already has. If someone says the rules of negation are not arbitrary because negation could not be such that $\sim\sim p = \sim p$, all that could be meant is that the latter rule would not correspond to the English word "negation". The objection that the rules are not arbitrary comes from the feeling that they are responsible to the meaning. But how is the meaning of "negation" defined, if not by the rules? $\sim\sim p = p$ does not follow from the meaning of "not" but constitutes it. Similarly, $p.p \supset q. \supset .q$ does not depend on the meanings of "and" and "implies"; it constitutes their meaning. If it is said that the rules of negation are not arbitrary inasmuch as they must not contradict each other, the reply is that if there were a contradiction among them we should simply no longer call certain of them rules. ["It is part of the grammar of the word 'rule' that if 'p' is a rule, '$p. \sim p$' is not a rule."*]

3 Logic proceeds from premises just as physics does. But the primitive propositions of physics are results of very general experience, while those of logic are not. To distinguish between the propositions of physics and those of logic, more must be done than to produce predicates such as *experiential* and *self-evident*. It must be shown that a grammatical rule holds for one and not for the other.

4 In what sense are laws of inference laws of thought?

Can a reason be given for thinking as we do? Will this require an answer outside the game of reasoning? There are two senses of "reason": reason for, and cause. These are two different orders of things. One needs to decide on a criterion for something's being a reason before reason and cause can be distinguished. Reasoning is the calculation actually done, and a reason goes back one step in the calculus. A reason is a reason only inside the game. To give a reason is to go

**Philosophische Grammatik*, Oxford and New York, 1969, p. 304.

through a process of calculation, and to ask for a reason is to ask how one arrived at the result. The chain of reasons comes to an end, that is, one cannot always give a reason for a reason. But this does not make the reasoning less valid. The answer to the question, Why are you frightened?, involves a hypothesis if a cause is given. But there is no hypothetical element in a calculation.

To do a thing for a certain reason may mean several things. When a person gives as his reason for entering a room that there is a lecture, how does one know that is his reason? The reason may be nothing more than just the one he gives when asked. Again, a reason may be the way one arrives at a conclusion, e.g., when one multiplies 13×25. It is a calculation, and is the justification for the result 325. The reason for fixing a date might consist in a man's going through a game of checking his diary and finding a free time. The reason here might be said to be included in the act he performs. A cause could not be included in this sense.

We are talking here of the grammar of the words "reason" and "cause": in what cases do we say we have given a reason for doing a certain thing, and in what cases, a cause? If one answers the question "Why did you move your arm?" by giving a behavioristic explanation, one has specified a cause. Causes may be discovered by experiments, but experiments do not produce reasons. The word "reason" is not used in connection with experimentation. It is senseless to say a reason is found by experiment. The alternative, "mathematical argument or experiential evidence?" corresponds to "reason or cause?"

5 Where the class defined by f can be given by an enumeration, i.e., by a list, $(x)fx$ is simply a logical product and $(\exists x)fx$ a logical sum. E.g., $(x)fx. = .fa.fb.fc$, and $(\exists x)fx. = .fa \vee fb \vee fc$. Examples are the class of primary colors and the class of tones of the octave. In such cases it is not necessary to add "and *a, b, c,* . . . are the only *f*'s". The statement, "In this picture I see all the primary colors", means "I see red and green and blue . . .", and to add "and these are all the primary colors" says neither more nor less than "I see all . . ."; whereas to add to "*a, b, c* are people in the room" that *a, b, c* are all the people in the room says more than "$(x)x$ is a person in the room", and to omit it is to say less. If it is correct to say the general proposition is a shorthand for a logical product or sum, as it is in some cases, then the class of things named in the product or sum is defined in the grammar, not by properties. For example, *being a tone of the octave* is

not a quality of a note. The tones of an octave are a list. Were the world composed of "individuals" which were given the names "*a*", "*b*", "*c*", etc., then, as in the case of the tones, there would be no proposition "and these are all the individuals".

Where a general proposition is a shorthand for a product, deduction of the special proposition fa from $(x)fx$ is straightforward. But where it is not, how does fa follow? "Following" is of a special sort, just as the logical product is of a special sort. And although $(\exists x)fx.fa. = .fa$ is analogous to $p \vee q.p. = .p$, fa "follows" in a different way in the two cases where $(\exists x)fx$ is a shorthand for a logical sum and where it is not. We have a different calculus where $(\exists x)fx$ is not a logical sum. fa is not deduced as p is deduced in the calculus of T's and F's from $p \vee q.p$. I once made a calculus in which *following* was the same in all cases. But this was a mistake.

Note that the dots in the disjunction $fa \vee fb \vee fc \vee \ldots$ have different grammars: (1) "and so on" indicates laziness when the disjunction is a shorthand for a logical sum, the class involved being given by an enumeration, (2) "and so on" is an entirely different sign with new rules when it does not correspond to any enumeration, e.g., "2 is even v 4 is even v 6 is even . . .", (3) "and so on" refers to positions in visual space, as contrasted with positions correlated with the numbers of the mathematical continuum. As an example of (3) consider "There is a circle in the square". Here it might appear that we have a logical sum whose terms could be determined by observation, that there is a number of positions a circle could occupy in visual space, and that their number could be determined by an experiment, say, by coordinating them with turns of a micrometer. But there is no number of positions in visual space, any more than there is a number of drops of rain which you see. The proper answer to the question, "How many drops did you see?", is *many,* not that there was a number but you don't know how many. Although there are twenty circles in the square, and the micrometer would give the number of positions coordinated with them, visually you may not see twenty.

6 I have pointed out two kinds of cases (1) those like "In this melody the composer used all the notes of the octave", all the notes being enumerable, (2) those like "All circles in the square have crosses". Russell's notation assumes that for every general proposition there are names which can be given in answer to the question "*Which* ones?" (in contrast to, "*What sort?*"). Consider $(\exists x)fx$, the notation for

"There are men on the island" and for "There is a circle in the square". Now in the case of human beings, where we use names, the question "Which men?" has meaning. But to say there is a circle in the square may not allow the question "Which?" since we have no names "*a*", "*b*", etc. for circles. In *some* cases it is senseless to ask "Which circle?", though "What sort of circle is in the square—a red one?, a large one?" may make sense. The questions "which?" and "What sort?" are muddled together [so that we think both always make sense].

Consider the reading Russell would give of his notation for "There is a circle in the square": "There is a thing which is a circle in the square". What is the *thing?* Some people might answer: the *patch* I am pointing to. But then how should we write "There are three patches"? What is the substrate for the property of being a patch? What does it mean to say "All things are circles in the square", or "There is not a thing that is a circle in the square" or "All patches are on the wall"? What are the *things?* These sentences have no meaning. To the question whether a meaning mightn't be given to "There is a thing which is a circle in the square" I would reply that one might mean by it that one out of a lot of shapes in the square was a circle. And "All patches are on the wall" might mean something if a contrast was being made with the statement that some patches were elsewhere.

7 What is it to look for a hidden contradiction, or for the proof that there is no contradiction? "To look for" has two different meanings in the phrases "to look for something at the North Pole", "to look for a solution to a problem". One difference between an expedition of discovery to the North Pole and an attempt to find a mathematical solution is that with the former it is possible to describe beforehand what is looked for, whereas in mathematics when you describe the solution you have made the expedition and have found what you looked for. The description of the proof is the proof itself, whereas to find the thing at the North Pole [it is not enough to describe it]. You must make the expedition. There is no meaning to saying you can describe beforehand what a solution will be like in mathematics except in the cases where there is a known method of solution. Equations, for example, belong to entirely different games according to the method of solving them.

To ask whether there is a hidden contradiction is to ask an ambiguous question. Its meaning will vary according as there is, or is not, a method of answering it. If we have no way of looking for it, then "contradiction" is not defined. In what sense could we describe it? We might seem to have fixed it by giving the result, $a \neq a$. But it is a result only if it is in organic connection with the construction. To find a contradiction is to construct it. If we have no means of hunting for a contradiction, then to say there might be one has no sense. We must not confuse what we can do with what the calculus can do.

8 Suppose the problem is to find the construction of a pentagon. The teacher gives the pupil the general idea of a pentagon by laying off lengths with a compass, and also shows the construction of triangles, squares, and hexagons. These figures are coordinated with the cardinal numbers. The pupil has the cardinal number 5, the idea of construction by ruler and compasses, and examples of constructions of regular figures, but not the law. Compare this with being taught to multiply. Were we taught all the results, or weren't we? We may not have been taught to do 61×175, but we do it according to the rule which we have been taught. Once the rule is known, a new instance is worked out easily. We are not given all the multiplications in the enumerative sense, but we are given *all* in one sense: any multiplication can be carried out according to rule. Given the law for multiplying, any multiplication can be done. Now in telling the pupil what a pentagon is and showing what constructions with ruler and compasses are, the teacher gives the appearance of having defined the problem entirely. But he has not, for the series of regular figures is a law, but not a law within which one can find the construction of the pentagon. When one does not know how to construct a pentagon one usually feels that the result is clear but the method of getting to it is not. But the result is not clear. The constructed pentagon is a new idea. It is something we have not had before. What misleads us is the similarity of the pentagon constructed to a measured pentagon. We call our construction the construction of the pentagon because of its similarity to a perceptually regular five-sided figure. The pentagon is analogous to other regular figures; but to tell a person to find a construction analogous to the constructions given him is not to give him any idea of the construction of a pentagon. Before the actual construction he does not have the idea of the construction.

When someone says there must be a law for the distribution of primes despite the fact that neither the law nor how to go about finding it is known, we feel that the person is right. It appeals to something in us. We take our idea of the distribution of primes from their distribution in a finite interval. Yet we have no clear idea of the distribution of primes. In the case of the distribution of even numbers we can show it thus: 1, **2,** 3, **4,** 5, **6,** . . . , and also by mentioning a law which we could write out algebraically. In the case of the distribution of primes we can only show: 1, **2, 3,** 4, **5,** 6, **7,** . . . [Finding a law would give a new idea of distribution] just as a new idea about the trisection of an angle is given when it is proved that it is not possible by straight edge and compasses. Finding a new method in mathematics changes the game. If one is given an idea of proof by being given a series of proofs, then to be asked for a new proof is to be asked for a new idea of proof.

Suppose someone laid off the points on a circle in order to show, as he imagined, the trisection of an angle. We would not be satisfied, which means that he did not have *our* idea of trisection. In order to lead him to admit that what he had was not trisection we should have to lead him to something new. Suppose we had a geometry allowing only the operation of bisection. The impossibility of trisection in this geometry is exactly like the impossibility of trisecting an angle in Euclidean geometry. And this geometry is not an incomplete Euclidean geometry.

9 Problems in mathematics are not comparable *in difficulty;* they are entirely *different* problems. Suppose one was told to prove that a set of axioms is free from contradiction but was supplied with no method of doing it. Or suppose it was said that someone had done it, or that he had found seven 7's in the development of π. Would this be understood? What would it mean to say that there is a proof that there are seven 7's but that there is no way of specifying where they are? Without a means of finding them the concept of π is the concept of a construction which has no connection with the idea of seven 7's. Now it does make sense to say *"There are seven 7's in the first 100 places"*, and although "There are seven 7's in the development" does not mean the same as the italicized sentence, one might maintain that it nevertheless makes sense since it follows from something which does make sense. Even though you accepted this as a rule, it is only *one* rule. I want to say that if you have a proof of the existence of seven 7's which does not tell you where they are, the sentence for the existence theorem has an entirely different meaning than one for which a means for finding them is given. To say that a contradiction is hid-

den, where there is nevertheless a way of finding it, makes sense, but what is the sense in saying there is a hidden contradiction when there is no way? Again, compare a proof that an algebraic equation of nth degree has n roots, in connection with which there is a method of approximation, with a proof for which no such method exists. Why call the latter a proof of *existence?*

Some existence proofs consist in exhibiting a particular mathematical structure, i.e., in "constructing an entity". If a proof does not do this, "existence proof" and "existence theorem" are being used in another sense. Each new proof in mathematics widens the meaning of "proof". With Fermat's theorem, for example, we do not know what it would be like for it to be proved.*

What "existence" means is determined by the proof. The end-result of a proof is not isolated from the proof but is like the end surface of a solid. It is organically connected with the proof which is its body.

In a construction as in a proof we *seem* first to give the result and then find the construction or proof. But one cannot point out the result of a construction without giving the construction. The construction is the end of one's efforts rather than a means to the result. The result, say a regular pentagon, only matters insofar as it is an incitement to make certain manipulations. It would not be useless. For example, a teacher who told someone to find a color beyond the rainbow would be expressing himself incorrectly, but what he said would have provided a useful incitement to the person who found ultra-violet.

10 If an atomic proposition is one which does not contain *and, or,* or apparent variables, then it might be said that it is not possible to distinguish atomic from molecular propositions. For p may be written as $p.p$ or $\sim\sim p$, and fa as $fa \vee fa$ or as $(\exists x)fx.x=a$. But "and", "or", and the apparent variables are so used that they can be eliminated from these expressions by the rules. So we can disregard these purportedly molecular expressions. The word "and", for example, is differently used in cases where it can be eliminated from those in which it cannot. Whether a proposition is atomic, i.e., whether it is not a truth-function of other propositions, is to be decided by applying certain methods of analysis laid down strictly. But when we have no method, it makes no sense to say there may be a hidden logical constant. The question whether such a seemingly atomic proposition as "It rains" is molecu-

*This paragraph is taken from The Yellow Book. (Editor)

lar, that it is, say, a logical product, is like asking whether there is a hidden contradiction when there is no method of answering the question. Our method might consist in looking up definitions. We might find that "It's rotten weather", for example, means "It is cold and damp". Having a means of analyzing a proposition is like having a method for finding out whether there is a 6 in the product 25×25, or like having a rule which allows one to see whether a proposition is tautologous.

Russell and I both expected to find the first elements, or "individuals", and thus the possible atomic propositions, by logical analysis. Russell thought that subject-predicate propositions, and 2-term relations, for example, would be the result of a final analysis. This exhibits a wrong idea of logical analysis: logical analysis is taken as being like chemical analysis. And we were at fault for giving no examples of atomic propositions or of individuals. We both in different ways pushed the question of examples aside. We should not have said "We can't give them because analysis has not gone far enough, but we'll get there in time". Atomic propositions are not the result of an analysis which has yet to be made. We can talk of atomic propositions if we mean those which on their face do not contain "and", "or", etc., or those which in accordance with methods of analysis laid down do not contain these. There are no hidden atomic propositions.

11 In teaching a child language by pointing to things and pronouncing the words for them, where does the use of a proposition start? If you teach him to touch certain colors when you say the word "red", you have evidently not taught him sentences. There is an ambiguity in the use of the word "proposition" which can be removed by making certain distinctions. I suggest defining it arbitrarily rather than trying to portray usage. What is called understanding a sentence is not very different from what a child does when he points to colors on hearing color words. Now there are all sorts of language-games suggested by the one in which color words are taught: games of orders and commands, of question and answer, of questions and "Yes" and "No." We might think that in teaching a child such language games we are not teaching him a language but are only preparing him for it. But these games are complete; nothing is lacking. It might be said that a child who brought me a book when I said "The book, please" would not understand this to mean "Bring me a book", as would an adult. But this full sentence is no more complete than "book". Of course

"book" is not what we call a sentence. A sentence in a language has a particular sort of jingle. But it is misleading to suppose that "book" is a shorthand for something longer which might be in a person's mind when it is understood. The word "book" might not lack anything, except to a person who had never heard elliptic sentences, in which case he would need a table with the ellipses on one side and sentences on the other.

Now what role do truth and falsity play in such language-games? In the game where the child responds by pointing to colors, truth and falsity do not come in. If the game consists in question and answer and the child responds, say, to the question "How many chairs?", by giving the number, again truth and falsity may not come in, though it might if the child were taught to reply "Six chairs agrees with reality". If he had been taught the use of "true" and "false" instead of "Yes" and "No", they would of course come in. Compare how differently the word "false" comes into the game where the child is taught to shout "red" when red appears and the game where he is to guess the weather, supposing now that we use the word "false" in the following circumstances: when he shouts "green" when something red appears, and when he makes a wrong guess about the weather. In the first case the child has not got hold of the game, he has offended against the rules; in the second he has made a mistake. The two are like playing chess in violation of the rules, and playing it and losing.

In a game where a child is taught to bring colors when you say "red", etc., you might say that "Bring me red" and "I wish you to bring me red" are equivalent to "red"; in fact that until the child understands "red" as information about the state of mind of the person ordering the color he does not understand it at all. But "I wish you to bring me red" adds nothing to *this* game. The order "red" cannot be said to describe a state of mind, e.g., a wish, unless it is part of a game containing descriptions of states of mind. "I wish . . ." is part of a larger game if there are two people who express wishes. The word "I" is then not replaceable by "John". A new multiplicity means having another game.

I have wanted to show by means of language-games the vague way in which we use "language", "proposition", "sentence". There are many things, such as orders, which we may or may not call propositions; and not only one game can be called language. Language-games are a clue to the understanding of logic. Since what we call a proposition is more or less arbitrary, what we call logic plays a different role from that which Russell and Frege supposed. We mean all sorts of things by "proposition", and it is wrong to start with a definition of a proposition and build up logic from that. If "proposition" is defined by reference to the notion of a truth-function, then arithmetic equations are also propositions—which does not make them the same as such a proposition as "He ran out of the building". When Frege tried to develop mathematics from logic he thought the calculus of logic was *the* calculus, so that what followed from it would be correct mathematics. Another idea on a par with this is that all mathematics could be derived from cardinal arithmetic. Mathematics and logic were one building, with logic the foundation. This I deny; Russell's calculus is one calculus among others. It is a bit of mathematics.

12 It was Frege's notion that certain words are unique, on a different level from others, e.g., "word", "proposition", "world". And I once thought that certain words could be distinguished according to their philosophical importance: "grammar", "logic", "mathematics". I should like to destroy this appearance of importance. How is it then that in my investigations certain words come up again and again? It is because I am concerned with language, with troubles arising from a particular use of language. The characteristic trouble we are dealing with is due to our using language automatically, without thinking about the rules of grammar. In general the sentences we are tempted to utter occur in practical situations. But then there is a different way we are tempted to utter sentences. This is when we look at language, consciously direct our attention on it. And then we make up sentences of which we say that they also ought to make sense. A sentence of this sort might not have any particular use, but because it sounds English we consider it sensible. Thus, for example, we talk of the flow of time and consider it sensible to talk of its flow, after the analogy of rivers.

13 If we look at a river in which numbered logs are floating, we can describe events on land with reference to these, e.g., "*When* the 105th log passed, I ate dinner". Suppose the log makes a bang on passing me. We can say these bangs are separated by equal, or unequal, intervals. We could also say one set of bangs was twice as fast as another set. But the equality or inequality of intervals so measured is entirely different from that measured by a clock. The phrase "length of interval" has its sense in virtue of the way we determine it, and differs according to the method of measurement. Hence the criteria for equal-

ity of intervals between passing logs and for equality of intervals measured by a clock are different. We cannot say that two bangs two seconds apart differ only in degree from those an hour apart, for we have no feeling of rhythm if the interval is an hour long. And to say that one rhythm of bangs is faster than another is different from saying that the interval between these two bangs passed much more slowly than the interval between another pair.

Suppose that the passing logs seem to be equal distances apart. We have an experience of what might be called the velocity of these (though not what is measured by a clock). Let us say the river moves uniformly in this sense. But if we say *time* passed more quickly between logs 1 and 100 than between logs 100 and 200, this is only an analogy; really nothing has passed more quickly. To say time passes more quickly, or that time flows, is to imagine *something* flowing. We then extend the simile and talk about the direction of time. When people talk of the direction of time, precisely the analogy of a river is before them. Of course a river can change its direction of flow, but one has a feeling of giddiness when one talks of time being reversed. The reason is that the notion of flowing, of *something,* and of the direction of the flow is embodied in our language.

Suppose that at certain intervals situations repeated themselves, and that someone said time was circular. Would this be right or wrong? Neither. It would only be another way of expression, and we could just as well talk of a circular time. However, the picture of time as flowing, as having a direction, is one that suggests itself very vigorously.

Suppose someone said that the river on which the logs float had a beginning and will have an end, that there will be 100 more logs and that will be the end. It might be said that there is an *experience* which would verify these statements. Compare this with saying that time ceases. What is the criterion for its ceasing or for its going on? You might say that time ceases when "Time River" ceases. Suppose we had no substantive "time", that we talked only of the passing of logs. Then we could have a measurement of time without any substantive "time". Or we could talk of time coming to an end, meaning that the logs came to an end. We could *in this sense* talk of time coming to an end.

Can time go on apart from events? What is the criterion for time involved in "Events began 100 years ago and time began 200 years ago"? Has time been created, or was the world created in time? These questions are asked after the analogy of "Has this chair been made?", and are like asking whether order has been created (a "before" and "after"). "Time" as a substantive is terribly misleading. We have got to make the rules of the game before we play it. Discussion of "the flow of time" shows how philosophical problems arise. Philosophical troubles are caused by not using language practically but by extending it on looking at it. We form sentences and then wonder what they can mean. Once conscious of "time" as a substantive, we ask then about the creation of time.

14 If I asked for a description of yesterday's doings and you gave me an account, this account could be verified. Suppose what you gave as an account of yesterday happened *tomorrow*. This is a possible state of affairs. Would you say you *remembered* the future? Or would you say instead that you remembered the past? Or are both statements senseless?

We have here two independent orders of events (1) the order of events in our memory. Call this memory time. (2) the order in which information is got by asking different people, 5—4—3 o'clock. Call this information time. In information time there will be past and future with respect to a particular day. And in memory time, with respect to an event, there will also be past and future. Now if you want to say that the order of information is memory time, you can. And if you are going to talk about both information and memory time, then you can say that you remember the past. If you remember that which *in information time* is future, you can say "I remember the future".

15 It is not *a priori* that the world becomes more and more disorganized with time. It is a matter of experience that disorganization comes at a later rather than an earlier time. It is imaginable, for example, that by stirring nuts and raisins in a tank of chocolate they become unshuffled. But it is not a matter of experience that equal distributions of nuts and raisins *must* occur when they are swished about. There is no experience of something necessarily happening. To say that if equal distribution does not occur there *must* be a difference in weight of the nuts and raisins, even though these have not been weighed, is to assume some other force to explain the unshuffling. We tend to say that there *must* be some explanation if equal distribution does not occur. Similarly, we say of a planet's observed eccentric behavior that there must be some planet attracting it. This is analogous to saying that if

two apples were added to two apples and we found three, one must have vanished. Or like saying that a die must fall on one of six sides. When the possibility of a die's falling on edge is excluded, and not because it is a matter of experience that it falls only on its sides, we have a statement which no experience will refute—a statement of grammar. Whenever we say that something *must* be the case we are using a norm of expression. Hertz said that wherever something did not obey his laws there must be invisible masses to account for it. This statement is not right or wrong, but may be practical or impractical. Hypotheses such as "invisible masses", "unconscious mental events" are norms of expression. They enter into language to enable us to say there *must* be causes. (They are like the hypothesis that the cause is proportional to the effect. If an explosion occurs when a ball is dropped, we say that some phenomenon must have occurred to make the cause proportional to the effect. On hunting for the phenomenon and not finding it, we say that it has merely not yet been found.) We believe we are dealing with a natural law *a priori,* whereas we are dealing with a norm of expression that we ourselves have fixed. Whenever we say that something must be the case we have given an indication of a rule for the regulation of our expression, as if one were to say "Everybody is really going to Paris. True, some don't get there, but all their movements are preliminary".

The statement that there must be a cause shows that we have got a rule of language. Whether all velocities can be accounted for by the assumption of invisible masses is a question of mathematics, or grammar, and is not to be settled by experience. It is settled beforehand. It is a question of the adopted norm of explanation. In a system of mechanics, for example, there is a system of causes, although there may be no causes in another system. A system could be made up in which we would use the expression "My breakdown had no causes". If we weighed a body on a balance and took the different readings several times over, we could either say that there is no such thing as absolutely accurate weighing *or* that each weighing is accurate but that the weight changes in an unaccountable manner. If we say we are not going to account for the changes, then we would have a system in which there are no causes. We ought not say that there are no causes in nature, but only that we have a system in which there are no causes. Determinism and indeterminism are properties of a system which are fixed arbitrarily.

16 We begin with the question whether the toothache someone else has is the same as the toothache I have. Is his toothache merely outward behavior? Or is it that he has the same as I am having now but that I don't know it since I can only say of another person that he is manifesting certain behavior? A series of questions arises about personal experience. Isn't it thinkable that I have a toothache in someone else's tooth? It might be argued that my having toothache requires my mouth. But the experience of *my* having toothache is the same wherever the tooth is that is aching, and whoever's mouth it is in. The locality of pain is not given by naming a possessor. Further, isn't it imaginable that I live all my life looking in a mirror, where I saw faces and did not know which was my face, nor how my mouth was distinguished from anyone else's? If this were in fact the case, would I say I had toothache in *my mouth?* In a mirror I could speak with someone else's mouth, in which case what would we call me? Isn't it thinkable that I change my body and that I would have a feeling correlated with someone's else's raising his arm?

The grammar of "having toothache" is very different from that of "having a piece of chalk", as is also the grammar of "I have toothache" from "Moore has toothache". The sense of "Moore has toothache" is given by the criterion for its truth. For a statement gets its sense from its verification. The use of the word "toothache" when I have toothache and when someone else has it belongs to different games. (To find out with what meaning a word is used, make *several* investigations. For example, the words "before" and "after" mean something different according as one depends on memory or on documents to establish the time of an event.) Since the criteria for "He has toothache" and "I have toothache" are so different, that is, since their verifications are of different sorts, I might seem to be denying that he has toothache. But I am not saying he really hasn't got it. Of course he has it: it isn't that he behaves as if he had it but really doesn't. For we have criteria for his really having it as against his simulating it. Nevertheless, it is felt that I should say that I do not know he has it.

Suppose I say that when he has toothache he has what I have, except that I know it indirectly in his case and directly in mine. This is wrong. Judging that he has toothache is not like judging that he has money but I just can't see his billfold. Suppose it is held that I must judge indirectly since I can't feel *his* ache. Now what sense is there to this? And what sense is there to "I can feel my ache"? It makes sense to say "His ache is worse than mine", but not to say "I feel *my* toothache" and "Two people can't have the same pain". Consider the statement that no two people can ever see the same sense datum. If being in the same position as another person were taken as the criterion for someone's seeing the same sense datum as he does, then one could imagine a person seeing the same datum, say, by seeing through someone's head. But if there is no criterion for seeing the same datum, then "I can't know that he sees what I see" does not make sense. We are likely to muddle statements of fact which are undisputed with grammatical statements. Statements of fact and grammatical statements are not to be confused.

The question whether someone else has what I have when I have toothache may be meaningless, though in an ordinary situation it might be a question of fact, and the answer, "He has not", a statement of fact. But the philosopher who says of someone else, "He has not got what I have", is not stating a fact.* He is not saying that in fact someone else has not got toothache. It might be the case that someone else has it. And the statement that he has it has the meaning given it, that is, whatever sense is given by the criterion. The difficulty lies in the grammar of "*having* toothache". Nonsense is produced by trying to express in a proposition something which belongs to the grammar of our language. By "I can't feel his toothache" is meant that I can't *try*. It is the character of the logical *cannot* that one can't try. Of course this doesn't get you far, as you can ask whether you can try to try. In the arguments of idealists and realists somewhere there always occur the words "can", "cannot", "must". No attempt is made to prove their doctrines by experience. The words "possibility" and "necessity" express part of grammar, although patterned after their analogy to "physical possibility" and "physical necessity".

Another way in which the grammars of "I have toothache" and "He has toothache" differ is that it does not make sense to say "I seem to have toothache", whereas it is sensible to say "He seems to have toothache". The statements "I have toothache" and "He has toothache" have different verifications; but "verification" does not have the same meaning in the two cases. The verification of my having toothache is having it. It makes no sense for me to answer the

* "It is particularly difficult to discover that an assertion that a metaphysician makes expresses discontentment with our grammar when the words of his assertion can also be used to state a fact of experience."—*The Blue Book,* Oxford and New York, 1969, pp. 56–7. (Editor)

question, "How do you know you have toothache?", by "I know it because I feel it". In fact there is something wrong with the question; and the answer is absurd. Likewise the answer, "I know it by inspection". The process of inspection is looking, not seeing. The statement, "I know it by looking", *could* be sensible, e.g., concentrating attention on one finger among several for a pain. But as we use the word "ache" it makes no sense to say that I look for it: I do not say I will find out whether I have toothache by tapping my teeth. Of "He has toothache" it is sensible to ask "How do you know?", and criteria can be given which cannot be given in one's own case. In one's own case it makes no sense to ask "How do I know?"

It might be thought that since my saying "He seems to have toothache" is sensible but not my saying a similar thing of myself, I could then go on to say "This is so for him but not for me". Is there then a private language I am referring to, which he cannot understand, and thus that he cannot understand my statement that I have toothache? If this is so, it is not a matter of experience that he cannot. He is prevented from understanding, not because of a mental shortcoming but by a fact of grammar. If a thing is *a priori* impossible, it is excluded from language.

Sometimes we introduce a sentence into our language without realizing that we have to show rules for its use. (By introducing a third king into a chess game we have done nothing until we have given rules for it.) How am I to persuade someone that "I feel *my* pain" does not make sense? If he insists that it does he would probably say "I make it a rule that it makes sense". This is like introducing a third king, and I then would raise many questions, for example, "Does it make sense to say I have toothache but don't feel it?" Suppose the reply was that it did. Then I could ask how one knows that one has it but does not feel it. Could one find this out by looking into a mirror and on finding a bad tooth know that one has a toothache? To show what sense a statement makes requires saying how it can be verified and what can be done with it. Just because a sentence is constructed after a model does not make it part of a game. We must provide a system of applications.

The question, "What is its verification?", is a good translation of "How can one know it?". Some people say that the question, "How can one know such a thing?", is irrelevant to the question, "What is the meaning?" But an answer gives the meaning by showing the relation of the proposition to other propositions. That is, it shows what it follows from and what follows from it. It gives the grammar of the proposition, which is what the question, "What would it be like for it to be true?", asks for. In physics, for example, we ask for the meaning of a statement in terms of its verification.

I have remarked that it makes no sense to say "I seem to have toothache", which presupposes that it makes sense to say I can, or cannot, doubt it. The use of the word "cannot" here is not at all like its use in "I cannot lift the scuttle". This brings us to the question: What is the criterion for a sentence making sense? Consider the answer, "It makes sense if it is constructed according to the rules of grammar". Then does this question mean anything: What must the rules be like to give it sense? If the rules of grammar are arbitrary, why not let the sentence make sense by altering the rules of grammar? Why not simply say "I make it a rule that this sentence makes sense"?

17 To say what rules of grammar make up a propositional game would require giving the characteristics of propositions, their grammar. We are thus led to the question, What is a proposition? I shall not try to give a general definition of "proposition", as it is impossible to do so. This is no more possible than it is to give a definition of the word "game". For any line we might draw would be arbitrary. Our way of talking about propositions is always in terms of specific examples, for we cannot talk about these more generally than about specific games. We could begin by giving examples such as the proposition "There is a circle on the blackboard 2 inches from the top and and 5 inches from the side". Let us represent this as "(2,5)". Now let us construct something that would be said to make no sense: "(2,5,7)". This would have to be explained (and you could give it sense), or else you could say it is a mistake or a joke. But if you say it makes no sense, you can explain why by explaining the game in which it has no use. Nonsense can look less and less like a sentence, less and less like a part of language. "Goodness is red" and "Mr. S came to today's redness" would be called nonsense, whereas we would never say a whistle was nonsense. An arrangement of chairs *could* be taken as a language, so that certain arrangements would be nonsense. Theoretically you could always say of a symbol that it makes sense, but if you did so you would be called upon to explain its sense, that is, to show the use you give it, how you operate with it. The words "non-

sense'' and ''sense'' get their meaning only in particular cases and may vary from case to case. We can still talk of *sense* without giving a clear meaning to ''sense'', just as we talk of winning or losing without the meaning of our terms being absolutely clear.

In philosophy we give rules of grammar wherever we encounter a difficulty. {To show what we do in philosophy I compare playing a game by rules and just playing about.}* We might feel that a complete logical analysis would give the complete grammar of a word. But there is no such thing as a completed grammar. However, giving a rule has a use if someone makes an opposite rule which we do not wish to follow. When we discover rules for the use of a known term we do not thereby complete our knowledge of its use, and we do not tell people how to use the term, as if they did not know how. Logical analysis is an antidote. Its importance is to stop the muddle someone makes on reflecting on words.

18 To return to the differing grammars of ''I have toothache'' and ''He has toothache'', which show up in the fact that the statements have different verifications and also in the fact that it is sensible to ask, in the latter case, ''How do I know this?'', but not in the former. The solipsist is right in implying that these two are on different levels. I have said that we confuse ''I have a piece of chalk'' and ''He has a piece of chalk'' with ''I have an ache'' and ''He has an ache''. In the case of the first pair the verifications are analogous, although not in the case of the second pair. The function ''*x* has toothache'' has various values, Smith, Jones, etc. But not *I*. *I* is in a class by itself. The word ''I'' does not refer to a possessor in sentences about having an experience, unlike its use in ''I have a cigar''. We could have a language from which ''I'' is omitted from sentences describing a personal experience. {Instead of saying ''I think'' or ''I have an ache'' one might say ''It thinks'' (like ''It rains''), and in place of ''I have an ache'', ''There is an ache here''. Under certain circumstances one might be strongly tempted to do away with the simple use of ''I''. We constantly judge a language from the standpoint of the language we are accustomed to, and hence we think we describe phenomena incompletely if we leave out personal pronouns. It is as though we had omit-

*I shall throughout use braces to indicate insertions from The Yellow Book. (Editor)

ted pointing to something, since the word "I" seems to point to a person. But we can leave out the word "I" and still describe the phenomenon formerly described. It is not the case that certain changes in our symbolism are really omissions. One symbolism is in fact as good as the next; no one symbolism is necessary.}

19 The solipsist who says "Only my experiences are real" is saying that it is *inconceivable* that experiences other than his own are real.* This is absurd if taken to be a statement of fact. Now {if it is logically impossible for another person to have toothache, it is equally so for me to have toothache. To the person who says "Only I have real toothache" the reply should be: "If only you can have real toothache, there is no sense in saying 'Only I have real toothache'. Either you don't need 'I' or you don't need 'real' . . . 'I' is no longer opposed to anything. You had much better say 'There is toothache'." The statement, "Only I have real toothache," either has a commonsense meaning, or, if it is a grammatical proposition, it is meant to be a statement of a rule. The solipsist wishes to say, "I should like to put, instead of the notation 'I have real toothache' 'There is toothache' ". What the solipsist wants is not a notation in which the ego has a monopoly, but one in which the ego vanishes.†}

[Were the solipsist to embody in his notation the restriction of the epithet "real" to what we should call his experiences and exclude "A has real toothache" (where A is not he), this would come to using "There is real toothache" instead of "Smith (the solipsist) has toothache".* {Getting into the solipsistic mood means not using the word "I" in describing a personal experience.} [Acceptance of such a change is tempting] because the description of a sensation does not contain a reference to either a person or a sense organ. Ask yourself, How do I, the person, come in? How, for example, does a person enter into the description of a visual sensation? If we describe the visual field, no person necessarily comes into it. We can say the visual field has certain internal properties, but its being *mine* is not essential to its description. That is, it is not an intrinsic property of a visual sensation, or a pain, to belong to someone. There will be no such thing as *my* image or someone else's. The locality of a pain has nothing to do with the person who has it: it is not given by naming a possessor. Nor

*See *The Blue Book,* p. 59. (Editor)

†Based on notes of The Yellow Book taken by Margaret Masterman. (Editor)

is a body or an organ of sight necessary to the description of the visual field. The same applies to the description of an auditory sensation. The truth of the proposition, "The noise is approaching my right ear", does not require the existence of a *physical* ear; it is a description of an auditory experience, the experience being logically independent of the existence of my ears. The audible phenomenon is in an auditory space, and the subject who hears has nothing to do with the human body. Similarly, we can talk of a toothache without there being any teeth, or of thinking without there being a head involved. Pains have a space to move in, as do auditory experiences and visual data. The idea that a visual field belongs essentially to an organ of sight or to a human body having this organ is not based on *what* is seen. It is based on such facts of experience as that closing one's lids is accompanied by an event in one's visual field, or the experience of raising one's arm towards one's eye. It is an experiential proposition that an eye sees. We can establish connections between a human body and a visual field which are very different from those we are accustomed to. It is imaginable that I should see with my body rather than with my eyes, or that I could see with someone else's eyes and have toothache in his tooth. If we had a tube to our eyes and looked into a mirror, the idea of a perceiving organ could be dispensed with. Were all human bodies seen in a mirror, with a loudspeaker making the sounds when mouths moved, the idea of an ego speaking and seeing would become very different.

20 [The solipsist does not go through with a notation from which either "I" or "real" is deleted.] He says "Only *my* experiences are real", or "Only I have real toothache", or "The only pain that is real is what I feel". This provokes someone to object that surely his pain is real. And this would not really refute the solipsist, any more than the realist refutes the idealist. The realist who kicks the stone is correct in saying it is real if he is using the word "real" as opposed to "not real". His rejoinder answers the question, "Is it real or hallucinatory?", but he does not refute the idealist who is not deterred by his objection. They still seem to disagree. Although the solipsist is right in treating "I have toothache" as being on a different level from "He has toothache", his statement that he has something that no one else has, and that of the person who denies it, are equally absurd. "Only my experiences are real" and "Everyone's experiences are real" are equally nonsensical.

21 Let us turn to a different task. What is the criterion for "This is *my body*"? There is a criterion for "This is my nose": the nose would be possessed by the body to which it is attached. There is a temptation to say there is a soul to which the body belongs and that my body is the body that belongs to me. Suppose that all bodies were seen in a mirror, so that all were on the same level. I could talk of A's nose and *my* nose in the same way. But if I singled out a body as mine, the grammar changes. Pointing to a mirror body and saying "This is *my* body" does not assert the same relation of possession between me and my body as is asserted by "This is A's nose" between A's body and A's nose. What is the criterion for one of the bodies being mine? It might be said that the body which moved when I had a certain feeling will be mine. (Recall that the "I" in "I have a feeling" does not denote a possessor.) Compare "Which of these is my body?" with "Which of these is A's body?", in which "my" is replaced by "A's". What is the criterion for the truth of the answer to the latter? There is a criterion for this, which in the case of the answer to "Which is mine?" there is not. If all bodies are seen in a mirror and the bodies themselves become transparent but the mirror images remain, my body will be where the mirror image is. And the criterion for something being my nose will be very different from its belonging to the body to which it is attached. In the mirror world, will deciding which body is mine be like deciding which body is A's? If the latter is decided by referring to a voice called "A" which is correlated to the body, then if I answer "Which is my body?" by referring to a voice called Wittgenstein, it will make no sense to ask which is my voice.

There are two kinds of use of the word "I" when it occurs in answer to the question "Who has toothache?". For the most part the answer "I" is a sign coming from a certain body. [If when people spoke, the sounds always came from a loudspeaker and the voices were alike, the word "I" would have no use at all: it would be absurd to say "I have toothache". The speakers could not be recognized by it.*] Although there is a sense in which answering "I" to the question, "Who has toothache?", makes a reference to a body, even to this body of mine, my answer to the question whether I have toothache is not made by reference to any *body*. I have no need of a criterion. My body and the toothache are independent. Thus one answer to the question "Who?" is made by reference to a body, and another seems not to be, and to be of a different kind.

*From the Yellow Book notes of Margaret Masterman. (Editor)

22 Let us turn to the view, which is connected with "All that is real is my experience", namely, solipsism of the present moment: "All that is real is the experience of the present moment". (Cf. Wm. James' remark "The present thought *is* the only thinker", which makes the subject of thinking equivalent to the experience.) {We may be inclined to make our language such that we will call only the present experience "experience". This will be a solipsistic language, but of course we must not make a solipsistic language without saying exactly what we mean by the word which in our old language meant "present".} Russell said that remembering cannot prove that what is remembered actually occurred, because the world might have sprung into existence five minutes ago, with acts of remembering intact. We could go on to say that it might have been created one minute ago, and finally, that it might have been created in the present moment. Were this latter the situation we should have the equivalent of "All that is real is the present moment". Now if it is possible to say the world was created five minutes ago, could it be said that the world perished five minutes ago? This would amount to saying that the only reality was five minutes ago.

Why does one feel tempted to say "The only reality is the present"? The temptation to say this is as strong as that of saying that only *my* experience is real. The person who says only the present is real because past and future are not here has before his mind the image of something moving.

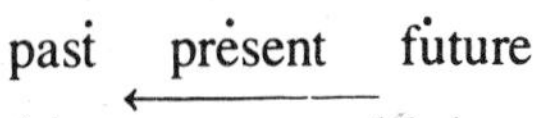

past present future
←————

This image is misleading, just as the blurred image we would draw of our visual field is misleading inasmuch as the field has no boundary. That the statement "Only the present experience is real" seems to mean something is due to familiar images we associate with it, images of things passing us in space. {When in philosophy we talk of the present, we seem to be referring to a sort of Euclidean point. Yet when we talk of present *experience* it is impossible to identify the present with such a point. The difficulty is with the word "present".} There is a grammatical confusion here. A person who says the present experience alone is real is not stating an empirical fact, comparable to the fact that Mr. S. always wears a brown suit. And the person who objects to the assertion that the present alone is real with "Surely the past and future are just as real" somehow does not meet the point. Both statements mean nothing.

By examining Russell's hypothesis that the world was created five minutes ago I shall try to explain what I mean in saying that it is meaningless. Russell's hypothesis was so arranged that nothing could bear it out or refute it. Whatever our experience might be, it would be in agreement with it. The point of saying that something has happened derives from there being a criterion for its truth. To lay down the evidence for what happened five minutes ago is like laying down rules for making measurements. The question as to what evidence there can be is a grammatical one. It concerns the sorts of actions and propositions which would verify the statement. It is a simple matter to make up a statement which will agree with experience because it is such that no proposition can refute it, e.g., "There is a white rabbit between two chairs whenever no observations or verifications are being carried out." Some people would say that this statement says more than "There is no white rabbit between the chairs", just as some would say it means something to say the world was created five minutes ago. When such statements are made they are somehow connected with a picture, say, a picture of creation. Hence it is that such sentences seem to mean something. But they are otiose, like wheels in a watch which have no function although they do not look to be useless.

I shall try to explain further what I mean by these sentences being meaningless by describing figures on two planes, one on plane I, which is to be projected, and the other, on plane II, the projection:

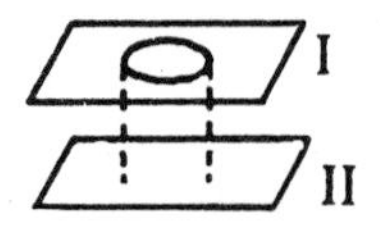

Now suppose the mode of projecting a circle on plane I was not orthogonal. In consequence, to say "There is a circle in plane II" would not be quite the same as saying that there is a circle in plane I. For a range of angles through which the circle is projected, the figures on plane II are all *more or less* circular. But now suppose the rays of light effecting the projection were allowed to vary through *any* range of angles. Then what meaning has it to say there are circles in plane II? When we give the method of projection such freedom, assertions about the projection become meaningless, though we still keep the picture of a circle in mind. Russell's assertion about the creation of the world is like this. The fact that there is a picture on plane I does not make a verifiable projection on plane II. We are accustomed to certain pictures being projected in a given way. But as soon as we leave this mode of projection, statements do not have their usual significance. When I say "That means nothing" I mean that you have altered your mode of projection. That it seems to mean something is due to an image of well-known things.

23 The words "thinkable" and "imaginable" have been used in comparable ways, what is imaginable being a special case of what is thinkable, e.g., a proposition and a picture. Now we can replace a visual image by a painted picture, and the picture can be described in words. Pictures and words are intertranslatable, for example,

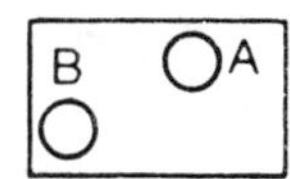

as A(5,7), B(2,3). A proposition is like, or something like, a picture. Let us limit ourselves to propositions describing the distribution of objects in a room. The distribution could be pictured in a painting. It would be sensible to say that a certain system of propositions corresponds to those painted and that other propositions do not correspond to pictures, for example, that someone whistles. Suppose we call the imaginable what can be painted, and the thinkable only what is imaginable. This would limit the word "thinkable" to the paintable. Now of course one can extend the way of picturing, for example, to someone whistling:

This is a new way of picturing, for a "rising" note is different from a vertical rise in space. With this new way we can imagine more, i.e., *think* more. People who make metaphysical assertions such as "Only the present is real" pretend to make a picture, as opposed to some other picture. I deny that they have done this. But how can I prove it? I cannot say "This is not a picture of anything, it is unthinkable" unless I assume that they and I have the same limitations on picturing. If I indicate a picture which the words suggest and they agree, then I can tell them they are misled, that the imagery in which they move does not lead them to such expressions. It cannot be denied that they have made a picture, but we can say they have been misled. We can say "It makes no sense in this system, and I believe this is the system you are using". If they reply by introducing a new system, then I have to acquiesce.

My method throughout is to point out mistakes in language. I am going to use the word "philosophy" for the activity of pointing out such mistakes. Why do I wish to call our present activity philosophy, when we also call Plato's activity philosophy? Perhaps because of a certain analogy between them, or perhaps because of the continuous development of the subject. Or the new activity may take the place of the old because it removes mental discomforts the old was supposed to.

24 With regard to a proposition about the external world or to a proposition of mathematics it is frequently asked "How do you know it?" There is an ambiguity here between reasons and causes. The interpretation we do *not* want is "How, causally, did you reach the result?" It does not matter what caused you to get the result; this is irrelevant. The important thing is to determine *what* you know when you are knowing it. To illustrate the distinction between reason and cause, let us take the question, How does one know the molecules of a gas are in motion? The answer might be psychological, for example, that you will see them if you have had enough to eat. If the kinetic theory were wrong, then no experience at all need correspond to it; but at the same time *there would be* a criterion for movement of molecules in a gas. The inventor of the theory would say "I am going to take such-and-such as a criterion". What is taken as a reason for belief in a theory is thus not a matter of experience but a matter of convention. If I believe the theory after taking clear soup, this is a cause of my belief, not a reason. When I am asked for a reason for the belief, what is expected, as part of the answer, is *what* I believe.

The different ways of verifying "It rained yesterday" help to determine the meaning. Now a distinction should be made between "being the meaning of" and "determining the meaning of". That I remember its raining yesterday helps determine the meaning of "It rained yesterday", but it is not true that "It rained yesterday" *means* "I remember that . . ." We can distinguish between primary and secondary criteria of its raining. If someone asks "What is rain?", you can point to rain falling, or pour some water from a watering can. These constitute primary criteria. Wet pavements constitute a secondary criterion and determine the meaning of "rain" in a less important way.

Two questions have been raised, which need to be answered now. (1) How could the meaning of a sentence about the past be given by a sentence about the present? (2) The verification of a proposition about the past is a set of propositions involving present and future tenses. If the verification gives the meaning, is part of the meaning left out? My reply is to deny that the verification *gives* the meaning. It merely *determines* the meaning, i.e., determines its use, or grammar.

25 When we understand a statement we often have certain characteristic experiences connected with it and with the words it contains. But the meaning of a symbol in our language is not the feelings it arouses nor the momentary impression it makes on us. The sense of a sentence is neither a succession of feelings nor one definite feeling. If you want to know the meaning of a sentence, ask for its verification. I stress the point that the meaning of a symbol is its place in the calculus, the way it is used. Of course if the symbol were used differently there might be a different feeling, but the feeling is not what concerns us. To know the meaning of a symbol is to know its use.

We can regard understanding a symbol, when we take its meaning in at a glance, as intuitive. Or understanding it may be discursive: knowing its meaning by knowing its use. Knowing the use of a sign is not a certain state lasting a certain time. (If we say knowing how to play chess is a certain state of mind, we have to say it is a hypothetical state.)

Attending to the way the meaning of a sentence is explained makes clear the connection between meaning and verification. Reading that Cambridge won the boat race, which verifies "Cambridge won", is obviously not the meaning, but it is connected with it. "Cambridge won" is not a disjunction, "I saw the race or I read the result or . . ." It is more complicated. Yet if we ruled out any one of the means of verifying the statement we would alter its meaning. It would upset our grammar if we excluded as a verification something that always accompanied winning. And if we did away with all means of verifying it we would destroy the meaning. It is clear that not every sort of verification is actually used to verify "Cambridge won", nor would just *any* verification give the meaning. The different verifications of the boat race being won have different places in the grammar of "boat race being won".

There is a mistaken conception of my view concerning the connection between meaning and verification which turns the view into idealism. This is that *a boat race = the idea of a boat race*. The mistake here is in trying to explain something in terms of something else. It lies back of Russell's definition of number, which we expect to tell us *what* a number *is*. The difficulty with these explanations in terms of something else is that the something else may have an entirely different grammar. Consider the word "chair". If there could be no visual picture of a chair, the word would have a different meaning. That one can see a chair is essential to the meaning of the word. But a visual picture of a chair is not a chair. What would it mean to sit on the visual picture of a chair? Of course we can explain what a chair is by showing pictures of it. But that does not mean that a chair is a complex of views. The tendency is to ask "What is a chair?"; but I ask how the word "chair" is used.

An intimately connected consideration concerns the words "time" and "length". People have felt that time is independent of the way it is measured. This is to forget what one would have to do to explain the word. Time is what is measured by a clock. To verify "The concert lasted an hour" you must tell how you measured time. It is a misunderstanding about both time and length that they are independent of measurement. If we have many ways of measuring which do not contradict, we do not assume any one way of measuring in explaining these words. The measuring which is connected with the meaning of a term is not exact, though in physics we do sometimes specify the temperature of the measuring rod. If, for example, we try to make the notion of a "precise time" more exact, we do not push it back far, for the striking of a clock at "precisely 4:30" takes time. And "to be *here* at precisely 4:30" is also not precise: should one be opening the door or be inside? Likewise with "having the same color". The verification of "These have the same color" may be that one can't see a color transition when they are put side by side, or that one can't tell the difference when they are apart, or that one can't tell one from the other when one is substituted for the other. These ways of testing give different meanings for "having the same color".

26 If the meaning of a word is determined by the rules for its use, does this mean that its meaning is the list of rules? No. Nor is the meaning, [as is sometimes the case with the bearer], *something* one can point to. The use of money and the use of words are analogous. Money is not always used to buy things which can be pointed to, e.g., when it buys permission to sit in a theatre, or a title, or one's life.

The ideas of meaning and sense are obsolete. Unless "sense" is used in such sentences as "This has no sense" or "This has the same sense as that", we are not concerned with sense.*

*This statement was not elaborated. See G.E. Moore's comment, *Philosophical Papers*, (London and New York, 1959), p. 258; first published in *Mind*, LXIII, 1954: "Wittgenstein's Lectures in 1930–33". (Editor)

In some cases it is not clear whether a statement is experiential or grammatical. How far is giving the verification of a proposition a grammatical statement about it? So far as it is, it can explain the meaning of its terms. Insofar as it is a matter of experience, as when one names a symptom, the meaning is not explained.

27 There is a problem connected with our talk of meaning: Does such talk indicate that I think meaning to be the subject matter of philosophy? Are we talking about something of more general importance than chairs, etc., so that we can take it that questions of meaning are the central questions of philosophy? Is meaning a meta-logical idea? No. For there are problems in philosophy that are not concerned with the meaning of "meaning", though perhaps with the meaning of other words, e.g., "time". The word "meaning" has no higher place than these. What gives it a different place is that our investigations are about language and about puzzles arising from the use of language. "Grammar", "proposition", "meaning" thus figure more often than other words, though investigation concerning the word "meaning" is on the same level as a grammatical investigation of the word "time".

Of course there isn't a philosophical grammar and ordinary English grammar, the former being more complete since it includes ostensive definitions such as the correlation of "white" with several of its applications, Russell's theory of descriptions, etc. These are not to be found in ordinary grammar books; but this is not the important difference. The important difference is in the aims for which the study of grammar are pursued by the linguist and the philosopher. One obvious difference is that the linguist is concerned with history, and with literary qualities, neither of which is of concern to us. Moreover, we construct languages of our own so as to solve certain puzzles which the grammarian is not interested in, e.g., puzzles arising from the expression "Time flows". We shall have to justify calling our comments on such a sentence grammar. If we say time flows in a different sense than water does, explaining this by an ostensive definition, we have indicated a way of explaining the word. And we have left the realm of what is generally called grammar. Our object is to get rid of certain puzzles. The grammarian has no interest in these; his aims and the philosopher's are different. We are pulling ordinary grammar to bits.

28 Let us look at the grammar of ethical terms, and such terms as "God", "soul", "mind", "concrete", "abstract". One of the chief troubles is that we take a substantive to correspond to a thing. Ordi-

nary grammar does not forbid our using a substantive as though it stood for a physical body. The words "soul" and "mind" have been used as though they stood for a thing, a gaseous thing. "What *is* the soul?" is a misleading question, as are questions about the words "concrete" and "abstract", which suggest an analogy with solid and gaseous instead of with a chair and the permission to sit on a chair. Another muddle consists in using the phrase "another *kind*" after the analogy of "a different *kind* of chair", e.g., that transfinite numbers are another kind of number than rationals, or unconscious thoughts a different kind of thought from conscious ones. The difference in the case of the latter pair is not analogous to that between a chair we see and a chair we don't see. The word "thought" is used differently when prefaced by these adjectives. What happens with the words "God" and "soul" is what happens with the word "number". Even though we give up explaining these words ostensively, by pointing, we don't give up explaining them in substantival terms. The reason people say that a number is a scratch on the blackboard is the desire to point to something. No sort of process of pointing is connected with explaining "number", any more than it is with explaining "permission to sit in a seat at the theatre".

Luther said that theology is the grammar of the word "God". I interpret this to mean that an investigation of the word would be a grammatical one. For example, people might dispute about how many arms God had, and someone might enter the dispute by denying that one could talk about arms of God. This would throw light on the use of the word. What is ridiculous or blasphemous also shows the grammar of the word.

29 Changing the meaning of a word, e.g., "Moses", when one is forced to give a different explication, does not indicate that it had no meaning before. The similarity between new and old uses of a word is like that between an exact and a blurred boundary. Our use of language is like playing a game according to the rules. Sometimes it is used automatically, sometimes one looks up the rules. Now we get into difficulties when we believe ourselves to be following a rule. We must examine to see whether we are. Do we use the word "game" to mean what all games have in common? It does not follow that we do, even though we were to find something they have in common. Nor is it true that there are discrete groups of things called "games". What is the reason for using the word "good"? Asking this is like asking why

one calls a given proposition a solution to a problem. It can be the case that one trouble gives way to another trouble, and that the resolution of the second difficulty is only connected with the first. For example, a person who tries to trisect an angle is led to another difficulty, posed by the question "Can it be done?" Proof of the impossibility of a trisection takes the place of the first investigation; the investigation has changed. When there is an argument about whether a thing is good, the discussion shows what we are talking about. In the course of the argument the word may begin to get a new grammar. In view of the way we have learned the word "good" it would be astonishing if it had a general meaning covering all of its applications. I am not saying it has four or five different meanings. It is used in different contexts because there is a transition between similar things called "good", a transition which continues, it may be, to things which bear no similarity to earlier members of the series. We *cannot* say "If we want to find out the meaning of 'good' let's find what all cases of good have in common". They may not have anything in common. The reason for using the word "good" is that there is a continuous transition from one group of things called good to another.

30 There is one type of explanation which I wish to criticize, arising from the tendency to explain a phenomenon by *one* cause, and then to try to show the phenomenon to be "really" another. This tendency is enormously strong. It is what is responsible for people saying that punishment must be one of three things, revenge, a deterrent, or improvement. This way of looking at things comes out in such questions as, Why do people hunt?, Why do they build high buildings? Other examples of it are the explanation of striking a table in a rage as a remnant of a time when people struck to kill, or of the burning of an effigy because of its likeness to human beings, who were once burnt. Frazer concludes that since people at one time were burnt, dressing up an effigy for burning is what remains of that practice. This may be so; but it need not be, for this reason. The idea which underlies this sort of method is that *every* time what is sought is *the* motive. People at one time thought it useful to kill a man, sacrifice him to the god of fertility, in order to produce good crops. But it is not true that something is always done because it is useful. At least this is not the sole reason. Destruction of an effigy may have its own complex of feelings without being connected with an ancient practice, or with usefulness. Simi-

larly, striking an object may merely be a natural reaction in rage. A tendency which has come into vogue with the modern sciences is to explain certain things by evolution. Darwin seemed to think that an emotion got its importance from one thing only, utility. A baby bares its teeth when angry because its ancestors did so to bite. Your hair stands on end when you are frightened because hair standing on end served some purpose for animals. The charm of this outlook is that it reduces importance to utility.

31 Let us change the topic to a discussion of *good*. One of the ways of looking at questions in ethics about *good* is to think that all things said to be good have something in common, just as there is a tendency to think that all things we call games have something in common. Plato's talk of looking for the essence of things was very like talk of looking for the *ingredients* in a mixture, as though qualities were ingredients of things. But to speak of a mixture, say of red and green colors, is not like speaking of a mixture of a paint which has red and green paints as ingredients. Suppose you say "*Good* is a quality of human actions and events". This is apparently an intelligible sentence. If I ask "How does one know an action has this quality?", you might tell me to examine it and I would find out. Now am I to investigate the movements making up the action, or are they only symptoms of goodness? If they are a symptom, then there must be some independent verification, otherwise the word "symptom" is meaningless. Now there is an important question which arises about goodness: Can one know an action in all its details and yet not know whether it is good? A similar question arises about beauty. Consider the beauty of a face. If all its shapes and colors are determined, is its beauty determined also? Or are these merely symptoms of beauty, which is to be determined otherwise? You may say that beauty is an indefinable quality, and that to say a particular face is beautiful comes to saying it has the indefinable quality. Is our scrutiny intended to find out whether a face has this indefinable quality, or merely to find out what the face is like? If the former, then the indefinable quality can be attributed to a particular arrangement of colors. But it need not be, and we must have some independent verification. If no separate investigation is required, then we only mean by a beautiful face a certain arrangement of colors and shapes.

32 The attribute beauty has been analyzed as what all beautiful things have in common. Consider one such property, agreeableness. I

call attention to the fact that in studying the laws of harmony in a harmony text there is no mention of "agreeableness"; psychology drops out. To say *Lear* is agreeable is to say something nondescriptive. And to many things this adjective is wholly inapplicable. Hence there is no basis for building up a calculus. The phrase "beautiful color", for example, can have a hundred meanings, depending on the occasion on which we use it.

Very often the adjectives we use are those applicable to the face of a person. This is the case with "beautiful" and "ugly". Consider how we learn such words. We do not as children discover the quality of beauty or ugliness in a *face* and find that these are qualities a *tree* has in common with it. The words "beautiful" and "ugly" are bound up with the words they modify, and when applied to a face are not the same as when applied to flowers and trees. We have in the latter a *similar* "game". For example, the adjective "stupid" is inapplicable to coals, except as you see a face in them. By a face being stupid we may mean it is the sort of face that really belongs to a stupid person; but usually not. Instead, it is a character of the particular expression of a face. This is not to say it is a character of the distribution of lines and colors. If it were, then one might ask how to find out whether the distribution is stupid. Is stupidity *part* of the distribution? The word "stupid" as applied to hands is still another game. The same is the case with "beautiful". It is bound up with a particular game. And similarly in ethics: the meaning of the word "good" is bound up with the act it modifies.

How can one know whether an action or event has the quality of goodness? And can one know the action in all of its details and not know whether it is good? That is, is its being good something that is independently experienced? Or does its being good follow from the thing's properties? If I want to know whether a rod is elastic I can find out by looking through a microscope to see the arrangement of its particles, the nature of their arrangement being a symptom of its elasticity, or inelasticity. Or I can test the rod empirically, e.g., see how far it can be pulled out. The question in ethics, about the goodness of an action, and in aesthetics, about the beauty of a face, is whether the characteristics of the action, the lines and colors of the face, are like the arrangement of particles: a *symptom* of goodness, or of beauty. Or do they constitute them? *a* cannot be a symptom of *b* unless there is a possible independent investigation of *b*. If no separate investigation is possible, then we mean by "beauty of face" a certain arrangement of colors and spaces. Now no arrangement is beautiful in itself. The word "beauty" is used for a thousand different things. Beauty of face is different from that of flowers and animals. That one is playing utterly different games is evident from the difference that emerges in the discussion of each. We can only ascertain the meaning of the word "beauty" by seeing how we use it.

33 What has been said of "beautiful" will apply to "good" in only a slightly different way. Questions which arise about the latter are analogous to those raised about beauty: whether beauty is inherent in an arrangement of colors and shapes, i.e., such that on describing the arrangement one would know it is beautiful, or not; or whether this arrangement is a symptom of beauty from which the thing's being beautiful is *concluded*.

In an actual aesthetic controversy or inquiry several questions arise: (1) How do we use such words as "beautiful"? (2) Are these inquiries psychological? Why are they so different, and what is their relation to psychology? (3) What features makes us say of a thing that it is the ideal, e.g., the ideal Greek profile?

Note that in an aesthetic controversy the word "beautiful" is scarcely ever used. A different sort of word crops up: "correct", "incorrect", "right", "wrong". We never say "This is beautiful enough". We only use it to say, "Look, how beautiful", that is, to call attention to something. The same thing holds for the word "good".

34 Why do we say certain changes bring a thing nearer to an ideal, e.g., making a door lower, or the bass in music quieter. It is not that we want in different cases to produce the same effect, namely, an agreeable feeling. What made the ideal Greek profile into an ideal, what quality? Actually what made us say it is the ideal is a certain very complicated role it played in the life of people. For example, the greatest sculptors used this form, people were taught it, Aristotle wrote on it. Suppose one said the ideal profile is the one occurring at the height of Greek art. What would this mean? The word "height" is ambiguous. To ask what "ideal" means is the same as asking what "height" and "decadence" mean. You would need to describe the instances of the ideal in a sort of serial grouping. And the word is always used in connection with one particular thing, for there is nothing in common between roast beef, Greek art, and German music. The word "decadence" cannot be explained without specific examples,

and will have different meanings in the case of poetry, music, and sculpture. To explain what decadence in music means you would need to discuss music in detail. The various arts have some analogy to each other, and it might be said that the element common to them is *the ideal*. But this is not the meaning of "the ideal". The ideal is got from a specific game, and can only be explained in some specific connection, e.g., Greek sculpture. There is no way of saying what all have in common, though of course one may be able to say what is common to two sculptures by studying them. In the statement that their beauty is what approaches the ideal, the word "ideal" is not used as is the word "water", which stands for something that can be pointed to. And no aesthetic investigation will supply you with a meaning of the word "ideal" which you did not have before.

When one describes changes made in a musical arrangement as being directed to bringing the arrangement of parts nearer to an ideal, the ideal is not before us like a straight line which is set before us when we try to draw it. (When questioned about what we are doing we might cite another tune which we thought *not* to be as near the ideal.) Some people say we have an ideal before our minds in the same way we have a memory image when we recognize a color. It *may* happen that you have a picture in mind with which the color recognized is compared, but this is rare. To see how the ideal comes in, say in making the bass quieter, look at what is being done and at one's being dissatisfied with the music as it is. Can one call this "action" of making the bass quieter an investigation? No, not in the sense of scientific investigation. No truth is found, except the psychological fact that I am satisfied with the result.

In what sense is aesthetic investigation a matter of psychology? The first thing we might say of a beautiful arrangement of colors—a flower, a meadow, or a face—is that it gives us pleasure. In saying these all give pleasure we speak as if the pleasure differed in degree rather than that the pleasures were of a different sort. Pain and pleasure do not belong on one scale, any more than the scale from boiling hot to ice cold is one of degree. They differ in kind. When a man jumps out of the window rather than meet the police he is not choosing the "*more* agreeable". Of course there are cases where we do weigh pleasures, as in choosing between cinemas. But this is not always the case. And it happens only sometimes that when we do not choose the lesser pain or the greater pleasure we choose what will produce these in the long run. One might think that it is entirely a matter of psychology whether something is good or beautiful, that in comparing musical arrangements, for example, one is making a psychological experiment to determine which produces the more pleasing effect. If this were true then the statement that beauty is what gives pleasure is an experiential one. But what people who say this wish to say is that it is not a matter of experience that beauty is what gives pleasure. Their statement is really a sort of tautology.

In aesthetic investigation the thing we are *not* interested in is *causal connections,* whereas in psychology we are. This is *the main point* of difference. To the question "*Why* is this beautiful?" we are accustomed to being satisfied with answers which cite *causes* instead of *reasons.* To name causal connections is to give an hypothesis. Giving a cause does not remove the aesthetic puzzle one feels when asked what makes a thing beautiful. It is useful to remind yourself of the answers given to the opposite question, "What is wrong with this poem or melody?", for the answer to the first question is of the same kind. The answer to "What is wrong with this melody?" is like the statement, "This is too loud", not like the statement that it produces sulphur in the blood.

The sort of experiment we carry on to discover people's likes and dislikes is not aesthetics. If it were, then you could say aesthetics is a matter of taste. In aesthetics the question is not "Do you like it?" but "Why do you like it?" Whenever we get to the point where the question is one of taste, it is no longer aesthetics. In aesthetic discussion what we are doing is more like solving a mathematical problem. It is not a psychological one. Aesthetic discussion is something that goes on inside the range of likes and dislikes. It goes on before any question of taste arises. A statement about a visual or auditory impression, as against what causes it, need not be psychological. That a sorrowful face becomes more sorrowful as the mouth turns downward is not a statement of psychology. In aesthetics we are not interested in causal connections but in description of a thing.

35 What is the justification for a feature in a work of art? I disagree with the answer "Something else would produce the wrong effect". Is it that you are satisfied, once something is found which removes the difficulty? What reasons can one give for being satisfied? The reasons are further descriptions. Aesthetics is descriptive. What it does is to *draw one's attention* to certain features, to place things side by side so as to exhibit these features. To tell a person "This is the climax" is

like saying "This is the man in the puzzle picture". Our attention is drawn to a certain feature, and from that point forward we see that feature. The reasons one gives for feeling satisfied have nothing to do with psychology. These, the aesthetic reasons, are given by placing things side by side, as in a court of law. If one gave *psychological* reasons for choosing a simile, those would not be reasons in aesthetics. They would be causes, not reasons. Stating a cause would be offering a hypothesis. Insofar as the remedy for the disagreeable feeling of top-heaviness of a door is like a remedy for a headache, a question concerning what remedy to prescribe is not a question of aesthetics. The aesthetic reason for feeling dissatisfied, as opposed to its cause, is not a proposition of psychology. A good example of a *cause* for dissatisfaction which I might have, say, with the way someone is playing a waltz, is that I have seen the waltz danced and know how it should be played. This does not give a reason for my dissatisfaction. The person who plays it, and I, have a different ideal of the waltz, and to give the reason for my dissatisfaction demands a *description*. Similarly, if a composition is felt to have a wrong ending.

36 I wish to remark on a certain sort of connection which Freud cites, between the fetal position and sleep, which looks to be a causal one but which is not, inasmuch as a psychological experiment cannot be made. His explanation does what aesthetics does: puts two factors together.

Another matter which Freud treats psychologically but whose investigation has the character of an aesthetic one is the nature of jokes. The question, "What is the nature of a joke?", is like the question, "What is the nature of a lyric poem?" I wish to examine in what way Freud's theory is a hypothesis and in what way not. The hypothetical part of his theory, the subconscious, is the part which is not satisfactory. Freud thinks it is part of the essential mechanism of a joke to conceal something, say, a desire to slander someone, and thereby to make it possible for the subconscious to express itself. He says that people who deny the subconscious really cannot cope with post-hypnotic suggestion, or with waking up at an unusual hour of one's own accord. When we laugh without knowing why, Freud claims that by psychoanalysis we can find out. I see a muddle here between a cause and a reason. Being clear why you laugh is not being clear about a *cause*. If it were, then agreement to the analysis given of the joke as explaining why you laugh would not be a means of detecting it. The success of the analysis is supposed to be shown by the person's agreement. There is nothing corresponding to this in physics. Of course we *can* give *causes* for our laughter, but whether those are in fact the causes is not shown by the person's agreeing that they are. A cause is found experimentally. The psychoanalytic way of finding why a person laughs is analogous to an aesthetic investigation. For the correctness of an aesthetic analysis must be agreement of the person to whom the analysis is given. The difference between a reason and a cause is brought out as follows: the investigation of a reason entails as an essential part one's agreement with it, whereas the investigation of a cause is carried out experimentally. ["What the patient agrees to can't be a *hypothesis* as to the *cause* of his laughter, but only that so-and-so was the *reason* why he laughed."*] Of course the person who agrees to the reason was not conscious at the time of its being his reason. But it is a way of speaking to say the reason was subconscious. It may be expedient to speak in this way, but the subconscious is a hypothetical entity which gets its meaning from the verifications these propositions have. What Freud says about the subconscious sounds like science, but in fact it is just *a means of representation.* New regions of the soul have not been discovered, as his writings suggest. The display of elements of a dream, for example, a hat (which may mean practically anything) is a display of similes. As in aesthetics, things are placed side by side so as to exhibit certain features. These throw light on our way of looking at a dream; they are reasons for the dream. [But his method of analyzing dreams is not analogous to a method for finding the causes of stomach-ache.†] It is a confusion to say that a reason is a cause seen from the inside. A cause is not seen from within or from without. It is found by experiment. [In enabling one to discover the reasons for laughter psychoanalysis provides] merely a representation of processes.

*G.E. Moore, "Wittgenstein's Lectures in 1930–33", *Philosophical Papers, p. 317. (Editor)*

†G.E. Moore, *ibid,* p. 316. (Editor)

PART II

The Yellow Book
(Selected Parts)

Wittgenstein's Lectures and Informal Discussions
during dictation of *The Blue Book*
1933–34

From the notes of Alice Ambrose

The Yellow Book

(Selected Parts)

1933–34

Lectures preceding dictation of *The Blue Book*

1 There is a truth in Schopenhauer's view that philosophy is an organism, and that a book on philosophy, with a beginning and end, is a sort of contradiction. One difficulty with philosophy is that we lack a synoptic view. We encounter the kind of difficulty we should have with the geography of a country for which we had no map, or else a map of isolated bits. The country we are talking about is language, and the geography its grammar. We can walk about the country quite well, but when forced to make a map, we go wrong. A map will show different roads through the same country, any one of which we can take, though not two, just as in philosophy we must take up problems one by one though in fact each problem leads to a multitude of others. We must wait until we come round to the starting point before we can proceed to another section, that is, before we can either treat of the problem we first attacked or proceed to another. In philosophy matters are not simple enough for us to say "Let's get a rough idea", for we do not know the country except by knowing the connections between the roads. So I suggest repetition as a means of surveying the connections. I shall begin by talking about problems connected with understanding, thinking, meaning. My investigation will not be psychological, even though a sentence is in a sense dead until it is understood. Before it is understood it is ink on paper. One might say it has meaning only for an understanding being. If there were no one to understand the signs we would not call the signs language.

2 The word "meaning" plays a great role in philosophy. Its importance is evident in discussions of the nature of mathematics. Frege ridiculed people for not seeing that the meaning of the signs "1",

"2", "3", etc. was the important thing, not the scratches on paper. It is a queer thing, however, that people have a propensity, on hearing the substantive "the number 1", to think of its meaning as being something beyond the sign and corresponding to it, in the way Smith corresponds to the name "Smith". There is of course a sense in which we can talk of the meaning but in which we cannot talk of the scratch. We use the word "one" in a way that we do not use the phrase "the sign 'one' ". It is, for example, nonsense to ask where the number 1 is. This comment may be trivial, like all the comments we shall make; but what is not trivial is seeing them all together.

It is useful to talk about chess, which is like mathematics but which has the virtue of having no nimbus like mathematics. [Both give rise to similar questions and similar remarks. "What is the king of chess?", "What is a number?"; "The rules of chess are about the king of chess, not about the wooden or ivory piece", "The rules of arithmetic are about numbers, not about signs on paper".] It is queer that when asked "What is the king of chess?" some people think of an ethereal entity as distinct from the piece. Similarly for "What is the number 1?" The question is misleading because, although it is correct to reply "There is no object corresponding to '1' in the sense that there is an object corresponding to 'Smith' ", we then look for an object in *another* sense. This is one of about a half-dozen traps we constantly fall into. When we hear the substantive word "number" used in the question "What is number?" our propensity is to think of an ethereal object. But what sort of answer can we give to this question? It is no use to say "Give a definition", for this gets us only one step further. As a way out of the difficulty posed by this question I suggest that we do not talk about the meaning of words but rather about the use of words. Suppose we take the meaning of a word to be *the way it is used*. To use the phrase "meaning of a word" as equivalent to "use of a word" has the advantage, among other things, of showing us something about the queer philosophical case where we talk of an object corresponding to the word. Normally we say an object corresponds to a word where in order to explain the word we point to an object, that is, give an ostensive definition. There are words whose meanings we can give by pointing to their bearers. Frege would say the object is the meaning. But "meaning" is not used in any such fashion. The phrase "meaning of a name" is not the same as "bearer of a name". The latter can be replaced by "Watson", but not the former. Obviously, "use of a word", if adopted as the definition of "meaning of a word", is not replaceable by "bearer of a word".

It might appear that we could give an ostensive definition of "Watson" but not of "1". But this is incorrect, as we can also give an ostensive definition of "1". Of course ostensive definitions differ. "Ostensive definition" is used in many different senses. The ostensive definition of "1" involves a different sort of pointing than the ostensive definition of "object", though one might point to the same thing in both cases. An ostensive definition is not really a definition at all. Ostensive definition is one rule only for the use of a word. And one rule is not enough to give the meaning. For example, from "This is sosh" you would not understand the use of the word "sosh", though from "This color is sosh" you would. That is, if a person is to learn the meaning of a word from such a definition he must already know what sort of thing it stands for. The word "color" already fixes the use of "sosh". The ostensive definition is of use if you need to fill in only one blank.

It has been suggested that *genus* plus *differentia* are equivalent to an ostensive definition. This is a prolific source of error. How are we to decide what the genus is? There is an inclination to believe that if a generic name is used for a number of things there must be something common to those things. It is a queer fact that things should have *one* generic name. It is a common belief that a definition of the generic name can give the common feature of the things the name is used for, for example, that what are called games all have something in common, which the definition of "game" can give. This notion is a trap. Our language is constructed on an apparently simple scheme, so that we are inclined to look at language as being much simpler than it is: we look for an object when we see a sign of the language; we think of anything we mention as falling under one genus only; and we look on the qualities of things as comparable to the ingredients of a mixture. It is difficult to avoid treating the genus, as common element, as if it were an ingredient which could be mixed in with other ingredients, because this notion is embodied in our language. But even if we had twelve liquids with one ingredient in common and these twelve had a generic name, it would not follow that the name was given because of this one ingredient. Games, for example, may not be called "games" because of a common element; there may be correlations merely between members of a series of games. And it might be that something is called a number which does not have anything in common with every sort of number but only with numbers of three sorts. Hence if you look for a justification for the use of a generic name you should not look for a quality all things it names have in common. A great muddle, for instance, resulted because people thought there was something in common between all the things called "good".

One important source of difficulty in philosophy is that words look so much alike. They are brought together in a dictionary like tools in a box, and like the tools, which look pretty much alike, they may have enormously different uses. The uses of words can differ from each other in the way beauty differs from a chair. They are incomparable in the way in which some things we buy are incomparable, such as a sofa and permission to sit in a theatre. When we talk of words and their meanings we tend to compare them with money and the things it buys rather than with money and the uses it has. A thing we buy with money is not the same as the use of the money, just as the bearer of a name is not the meaning of the name.

To revert to ostensive definition. I have said that it can only be understood if it makes the last decision about the word's use, that is, if it supplements a knowledge of the grammar of a word which is lacking one rule. There is no reason why you should not say that an ostensive definition fixes the differentia if the rest is known, provided you do not think there is but one genus and but one way of fixing the genus. But it is misleading and in a sense entirely false to say it can be understood only if it makes the last decision. For instance, as children we did not learn a rule for the use of "water" when water was pointed out, or know other rules of which this was the last rule. Of course we perhaps need not call this ostensive definition, but there is no clear line between ostensive definition for children and ostensive definition for grown-ups. There are stages in children's learning before which they cannot ask "What is this?", and even when they reach this stage they may yet not be able to ask "What color is this?" To describe ostensive definition we could give a number of games, distinguished as follows: (1) giving the last of a list of rules, (2) doing what children do when they learn the application of a word, (3) gradations between (1) and (2).

3 I have remarked that we are inclined to view our language as much simpler than it is. Cf. Augustine, who said that he learned Latin by learning the names of things. Surely he learned also such words as "not", "or", etc. We can criticize his view in either of two ways: that it is wrong, or that it describes a simpler thing than we call language. The latter may be compared to giving a description of games which applies only to a special class of games. Inasmuch as our lan-

guage is complex, I shall point out simpler structures which can be set side by side with it to see what light they shed on it.

Suppose a person learned a language by having people tell him the names of things after pointing to them. And suppose the language served one purpose only, say, for building a house with different shaped materials. The orders in which the names of these materials were called out would give the way in which the house was to be built. This would be a complete language. With language games such as this there is no standard of completeness, but we may as well say it is complete since we cannot say merely by looking at it that something is lacking. What we are doing here is like taking chess and making a simpler game involving simpler operations and a smaller number of pawns. In a sense simpler languages lead up to more complex ones, but the simpler ones are not incomplete.

Suppose that after saying something about Moses someone asked who he was, and he was *defined* as the man who led the Israelites out of Egypt. Suppose the objection was made that researches showed he had not done so, and another definition was offered, which in turn might be disputed. The changing definitions show that when the discussion began there was no definition of "Moses", that is, an exact game was not being played. At the same time, we would never say that nothing was meant, for there is a certain range of definitions from which one chooses.

It might be said of me that I describe language as if it were in a vacuum, but this is not so. What I do is to talk of language as consisting of fixed rules, which is really contrary to fact.

Consider the way we play a game and the way the rules enter into playing it. There could be a table of rules which we read, or which we know by heart and call up when playing, or we could play automatically. It is the same with the use of a definition. Suppose a definition of "leaf" were first given ostensively

and one were asked whether a shape not provided for in the definition was a leaf, say

Now can one tell where the line is to be drawn between rules one must know in order to understand what a leaf is and those which are not absolutely necessary? Suppose one said a leaf was anything falling within this general shape:

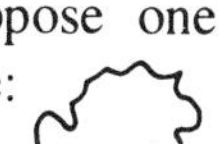

This could serve as a rule. But when we *use* a word without strict rules and later lay down strict rules for its use, its grammar cannot be entirely like that of its former use. It would be similar in the way a figure drawn with sharp outlines and a blurred figure are similar.

We shall compare the use of language to playing a game according to exact rules, because all philosophical troubles arise from making up too simple a system of rules. Philosophers try to tabulate the rules, and because there are so many things to mislead them, for instance, analogies, they lay down the rules wrongly. The only way to correct a wrong rule is to give another rule or set of rules according to which they do play. It is necessary to emphasize this because in discussing understanding, meaning, etc. our greatest difficulty is with the entirely fluid use of words. I shall not proceed by enumerating different meanings of the words "understanding", "meaning", etc., but instead shall draw ten or twelve pictures that are similar in some ways to the actual use of these words. My being able to draw these pictures is not because they all have something in common; their relationship may be quite complicated.

To begin with, I have suggested substituting for "meaning of a word", "use of a word", because *use of a word* comprises a large part of what is meant by "the meaning of a word". Understanding a word will thus come to knowing its use, its applications. The use of a word is what is defined by the rules, just as the use of the king of chess is defined by the rules. And just as the shape and material of the king of chess are irrelevant to its use, so are the shape and sound of a word to its use.

I also suggest examining the correlate expression "explanation of meaning". This will teach us something about the meaning of "meaning". Whereas it may be difficult to explain what "length" means, but not difficult to explain "measurement of length", analogously it is less difficult to describe what we call "explanation of meaning" than to explain "meaning". The meaning of a word is explained by describing its use.

It is a queer thing that, considering language as a game, the use of a word is internal to the game whereas its meaning seems to point to something outside the game. What seems to be indicated is that "meaning" and "use" are not equatable. But this is misleading.

4 Let us accept "understanding a word" as being knowledge of its use. It is useful to compare analogous questions about understanding a

word and playing chess: How do you know that you understand a word, e.g., "red"? How do you know that you are playing chess and not draughts? One reply is that you would not know it unless you had made a move which decided between them, or had made a wrong move in the game you claimed to be playing. If in knowing one is playing chess, or understanding a word, the rules went through one's head, then one would be entitled to claim that one knows this. But the rules do not go through one's head. The criteria for knowing that you are playing chess differ. One criterion would be giving the rules. But if the rules are not offered as criterion, then what? What one usually says is that whether one is playing chess, or understands a word, is assured by knowing one's intentions. But how do you know your intentions? Is it a fact that a particular psychological process exists, corresponding to a particular game? Is this something known by experience? The answer is that it is absurd to ask whether one knows one has a certain intention. And the same is true of wishing, thinking, hoping.

There may be a sense of "understanding" in which the word refers to a state of mind which occurs while making a move in chess or while using a word. In this connection compare two people, one of whom moves the pieces mechanically on a board, and the other who moves them with understanding. But there is also a sense in which "understanding a word" means knowing its use. The latter is very different from having a state of mind, although the two may be causally connected. There may be states of mind corresponding to every game, but these states do not presuppose or contain the rules.

It has been argued that if when one knows the use of a word one knows the rules, then one has the capacity to produce them on demand. This capacity might be considered a psychological state. The question then arises, What becomes of the distinction I have made between states of mind and knowledge of the rules? My reply is that "psychological state" is ambiguous. The distinction between a psychological state, meaning a capacity to produce rules on demand, and a psychological state, meaning a particular feeling, has a parallel in the distinction between subconscious and conscious states. If you object that knowing the use of a word, as well as a mental process accompanying the hearing or pronouncing of a word, is a state of mind, then you should distinguish between states of consciousness and states in the *hypothetical* sense. Knowing the alphabet, or the rules of chess, or the use of a word, is not a state of consciousness. To see that it is not, ask yourself what it is like to know the alphabet all the time. The grammar of the words "knowing the alphabet" and "being able to play chess" is entirely different from the grammar of the words "feeling something when you move a chessman". We can say this: that "understanding a word" is certainly used in two ways, for an accompanying mental process, and for knowing the use of the word. The grammars of "feeling something when we hear the word" and "knowing the word's use" are entirely different. To see how they differ, consider the parallel case of knowing the rules of chess.

Now you may question whether my constantly giving examples and speaking in similes is profitable. My reason is that parallel cases change our outlook because they destroy the uniqueness of the case at hand. For example, the Copernican revolution destroyed the idea that the earth has a unique place in the solar system. Let us turn then to the parallel between playing chess and understanding a word, and contrast the grammar of the words "knowing the rules of chess" with the grammar of "having a certain feeling while playing chess with understanding". It is important to note that the moves of an automaton are different from the same moves made consciously. States occur when we play with knowledge of the rules which do not occur otherwise. And yet knowing the rules is not a state of consciousness. For example, knowing the application of the word "and" is not the same as the "*and* feeling" of which William James spoke. And knowing the use of the verb "to be" in "The rose is red" and in "2 + 2 are 4" is different from the mental event corresponding to each occurrence of it. There is a tendency to suppose that we can swallow the meaning of a word as a whole whenever we understand it.

5 Consider the following analogy: between a cube or pyramid with one painted surface, behind which is an invisible body, and a word and the meaning behind it. Any position in which this surface could be placed will depend on the position of the solid body back of it. We are tempted to think that if we know a cube is back of the painted surface we can know the rules for arranging the surface with other surfaces. But this is not true. One cannot deduce the geometry of the cube from looking at the cube. Rules do not follow from an act of comprehension. Analogously, we are tempted to think we can deduce the rules for the use of a word from its meaning, which we supposedly grasp as a whole when we pronounce the word. This is the error I would eradicate. The difficulty is that inasmuch as we grasp the meaning without

grasping all the rules, it seems as if the rules *could* be *developed* from the meaning.

To say that the use of a word, e.g., "cube", follows from its meaning is to treat the word as if it were the visible face of a hidden body, its meaning, whose rules of combination with other hidden bodies are given by the laws of geometry. Could the geometry of cubes be deduced from the figures? Does geometry talk about cubes? It evidently does not talk about iron or copper cubes; but it might be claimed that it talks about geometrical cubes. In fact, geometry does not treat of cubes but of the grammar of the word "cube", as arithmetic treats of the grammar of numbers. The word "cube" is defined in a geometry, and a definition is not a proposition about a thing. If we alter the geometry we alter the meaning of the words used, for the geometry constitutes the meaning. If 457 were multiplied by 63 and a different result was got than in the ordinary game, this would mean that "cardinal number" is used with a different meaning. Arithmetical propositions say nothing about numbers, but determine which propositions about numbers make sense and which do not. Similarly, geometrical propositions say nothing about cubes, but determine which propositions about cubes make sense and which do not. This comment suggests the relation between mathematics and its application, i.e., between a sentence giving the grammar of a word and an ordinary sentence in which the word figures.

What role does a cube play in the geometry of a cube and in the development of that geometry? To answer this question we must distinguish between two sorts of investigations: investigation into the properties of an object and investigation into the grammar for the use of a word referring to the object. I want to say that a geometrical investigation, in the sense of an investigation into the properties of geometrical straight lines and cubes, is not possible. There is one sort of mistake that it is important to look at because of its pervasiveness. This is that the real cube and the geometrical cube are comparable. Geometry is not a physics of *geometrical* straight lines and cubes. It constitutes the meaning of the words "line" and "cube". The role the cube plays in its geometry is the role of a *symbol,* not that of a solid with which inaccurate real cubes are comparable. Figures like this one

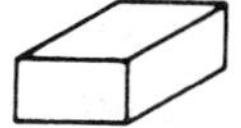

are part of the language of geometry and play the role of a symbol within geometrical proofs. It is for this reason that it does not matter whether a drawing is accurate.

6 It might be thought that if a mental act accompanying hearing or saying a word cannot sum up the meaning of the word in the sense in which rules define it, the mental act loses its importance. But it has importance in that sometimes, e.g., in understanding the word "red", it is essential to have an image before one, as when one is ordered to copy the particular red of this book. Here the word "red" alone is not enough. In such a case the image plus the word will function as a complete symbol, beyond which we would need nothing else. Remember, however, that in many cases a private mental act, which livens up the symbol like the soul a body, is not necessary. Instead of imagining red, one could in some cases use a sample red patch. There is no reason for supposing that if a red image is at all essential to thinking of red, an imagined thing is better than a seen thing. It is a prejudice to suppose that one must call up images in thinking.

One could say that in thinking one calculates with words and images. The calculation proceeds from one step to the other without any one step (mental act) containing the others. And there is no mental act anticipating the steps actually taken. So let us do away with mental acts in the description of thinking and simply talk of the calculus. Thinking is not something that accompanies talking; it may just be the talking. Some people have the idea that in a language, words follow the order of thinking. Does this mean that there is a separate process going alongside the words?

Compare the two questions, "Did you mean what you said?," to which the answer is Yes, or No, and "What did you mean?", to which the answer is another expression. Thus, in these two questions we have two uses of "mean". Now what is it to mean what one says? All sorts of things may justify your saying you mean what you say, but none of these needs to be a mental process accompanying the words.

In which cases would you say you are thinking while reading? It might be when you are having images, or it might be when you are able to write afterwards what you have read, regardless of having images. Understanding the sentences read may be a number of things, such as being attentive, remembering, or looking at a bit of paper of the required color when asked to copy a shade of red. We must not confuse the personal experience of remembering with a hypothetical act of mentally recording, supposedly done when you remember what you have read. The difficulties arising when we think about thinking, wishing, etc. have one main source, the tendency to find one process corresponding to the words "wishing" and "thinking" occurring in the expression of wishes and thoughts, comparable to the physical activity corresponding to the word "writes" in "He writes letters often". What we call wishing is not one *activity* hidden in all cases of wishing. It is not one process, like writing and speaking, and questions arise about wishing which are not present with the sentence "He writes . . ." What is it, for example, to wish that Smith would turn up? Can one wish throughout a certain time? The word "intention" presents the same obscurity.

In some cases it is nonsense to ask "Are you sure you wish it?", but there is a *hypothetical* use of the word "wish" in which it makes sense to ask "Are you sure?" In a case where one was not certain whether one wished, one way to go about finding out would be to ask what sort of thing *confirms* "I did not wish that". Flipping a coin can determine what you wished, and past experience may have taught you that an apple is what satisfies your hunger. By such means you find out whether the hypothesis that you wish for so-and-so is correct. There are of course wishes in which one has one definite feeling, others where there is a mixture of feelings, and others again where there is no definite feeling. The feelings accompanying wishing are very vague and coarse, not localized; or if localized, are organic. (In wishing for a pear, have you the same feeling as in wishing for an apple?) "I want water now" may be said by a person with a corresponding feeling of thirst; but the words "I shall want water later" will probably have no such feeling corresponding to them, and the feeling which does correspond, granted one exists, will be hard to describe.

What has been said here about the uses of the word "wishing" applies also to "meaning" and "interpreting": no one particular feeling accompanies them. Nor is understanding a sentence necessarily a sort of following of the sentence by imagining, though it sometimes happens that such a process accompanies the spoken or written sentence. Sometimes there is an amorphous feeing which cannot be translated into a sentence. But this is not always the case, since sometimes the expression of the thought is the thought. For example, sometimes the sentence "I expect Mr. Smith" *is* the expectation. Where wishing for a specific thing is a certain process, one can look at the process and see what is wished for. Here there could be no such question as "Are you sure *that* is what you wish?"

7 As for wishing or understanding being merely the expression of the wish or the thought, the ordinary objection to it is that no mere sign is

the thought; the thought interprets the sign. Thinking is not speaking or reading the symbols. Such an objection is rooted in the view that thinking, or some process in the mind, accompanies the symbols. Now is this supposed process something amorphous, a state having duration while the sentence is said, written, or heard? Perhaps it is something articulate, so that understanding a sentence consists of a series of interpretations, one interpretation for each word. This process would be translatable into a sentence, so that we could derive the sentence from the process or the process from the sentence. But this only adds one phenomenon to another.

That *pure thought* is conveyed by words and is something different from the words is a *superstition*. We are simply misled if we suppose that a symbol must first convey something else, say a picture, and that when an order is given one acts by interpreting the picture. Suppose an order to go in a certain direction is given. Calling up an image of this direction, say

would be an interpretation. But this interpretation is not necessary, for if one can make this interpretation, why not act on the words? The fact that in two different languages the thought expressed by a sentence is the same does not mean that one may go looking for the thought conveyed by them. *Where* is the thought? This question can be answered if "where" is interpreted. It is in some respects like the question, "Where is the individual's visual space?" There is no "where". It makes no sense to ask this. If on piercing a nerve the visual field is blotted out but returns when the piercing ceases, one could say the field is situated in this part if one knows what is meant by "situated". Specification of locality can be entirely different things. In a sense one might say the "where" of a thought is in the head, but in no important sense.

Now is there a good reason to oppose the process of thinking to the process of speaking? We are accustomed to saying we are having difficulty in *expressing* thought. What happens in this situation? Sometimes we have an image, but we may do many different things, e.g., make a gesture until the word comes. Likewise, there are lots of different processes we call "looking in our memory". The latter phrase is a simile taken from "looking in a room". Obviously looking in a room is different from looking in memory. There is a possibility of covering the area in the case of the former so that if what is sought is there one will find it. Also, we can say of looking in a room that the thing sought is either there or not. But this cannot be said of memory. Looking in memory is comparable to depending on a mechanism which either does or does not work, like pushing a row of buttons, none of which may ring the bell.

I repeat the point that the fact that two sentences express the same thought does not mean that there is a thing which is the thought, a gaseous being corresponding to the sentences. But we must not thereupon conclude that the word "thought" as contrasted with "sentence" does not mean anything. The two words have different uses, just as "king" and "king piece" have different uses. And just as with the comma we should not say "Here is the comma and there its meaning", so it is with the word "word". It has its function—its use.

8 To turn to a connected topic, voluntary and involuntary movement. What is the difference between them? Some would say it is the presence of a feeling. But a feeling may not be an accompaniment of a voluntary act, so this does not serve to distinguish one from the other. When one wills an action, what is the object of willing—the object one sees, or the contraction of a muscle? We must note that *willing* and *wishing* are entirely different. When I say I willed to raise my arm, I do not mean that I merely wished it very strongly and then the arm rose. Willing is not a thing which happens to me; it is a thing I do. The word "wish" has a much wider use than "will". The word "willing" is used in connection with phenomena bound up with our bodies. Thinking, by contrast with willing, is something which happens to one, not something one does.

The Yellow Book

(Selected Parts)

1933–34

Lectures and Informal Discussions in the intervals between dictation of *The Blue Book*

9 Difficulties in philosophy constantly occur in cases where there is claimed to be a special state of mind for which a word stands. The further one goes from states of mind to activities, say, the simpler the philosophical difficulties become. In talking of knowing, or of remembering, I shall therefore be interested in the meaning of "knowing" or "remembering" which comes as near as possible to meaning a particular state of mind or a number of states of mind. It must be emphasized, however, that it is not one particular state of mind that is involved in knowing, and the same for remembering. The activity of looking into one's memory for this morning's events is very peculiar. It is clearly different from remembering last night's events. Aristotle claimed that when we think of the future we look up, and when we think of the past, we look down. And it may quite well be that remembering *consists* in part in the position of one's muscles, or a feeling in one's neck. (Compare with William James' observation that "we are sad because we weep", that weeping is not an inessential accompaniment of an amorphous state.) What happens when I remember my toothache? Perhaps what happens is only that I say I remember it, though there is usually some sort of accompaniment.

Different sorts of memories are to be distinguished. One kind passes in time, cinematographically. Another is like an image given all at once, but afar off. And we must not fail to take account of the kind of memory which consists in remembering a poem or tune rather than some event of the past. In these cases "to remember it" means "to be able to reproduce it". In remembering a poem we do not first visualize the printed poem and then say it. We simply start off saying it, and the puzzling thing is the lack of any transition. If I am prepared to sing

"God Save the King," certainly all of the words do not pass through my head before I begin to sing, and at most only a fraction of them. But then what is the difference between being willing to sing "God Save the King" and being willing to sing "Deutschland, Deutschland über Alles"? The difference could be (1) that when asked, "Are you willing to sing 'Deutschland, . . .'?" you reply Yes, (2) that you will to do it, (3) that you sing it.

Like thinking, wishing, remembering, etc., being willing to do a certain action A is often thought to be a particular state of mind. And the same sort of questions arise: What has being willing to do *A* to do with *A?* What is the connection between the state of mind and the action? Is it an empirical one? In being willing to sing "*A*" you must know *what* you are willing to do inasmuch as there is no further evidence such as "This state of mind is often followed by '*A*' ". Suppose we use the words "being willing" in a derivative sense to mean a certain state of the muscles. When one is willing in this sense to sing "*A,*" *what* one is willing to do is a matter of experience; for this can be determined by experiment. Being willing and what one is willing to do are connected empirically. But in the ordinary sense, willingness and what one is willing to do are not so connected. If they were, it would make sense to ask "How do you know you are willing to do *A?*"

If "being willing" is to be considered a state of consciousness, and if you want it not to make sense to ask "Are you sure you are willing to sing '*A*'?", then knowing that you are willing must consist in somehow reading off from your willingness what you are willing to do. If there is any transition between willingness and what you are willing to do it would seem to be just this reading off from your willingness. Yet when we look at what happens when we are willing to do an act, a connecting link between being willing to do it and doing it seems to be lacking. This absence of transition is puzzling. We feel that because a link is lacking we are behaving like an automaton. By contrast, of a living being who says he is willing we have the idea that the distinguishing characteristic is that he makes up his mind to sing "*A*", remembers "*A*", and then sings it. The picture we have of being willing to do a thing is one in which making up our minds is one definite action. Will we allow such an act to be empirically connected with what happens? No. What we want is an action in which what we are going to do is already *performed*. That is, willingness should contain the action—being willing to sing aloud should be like singing to oneself. Here there is still a transition to be made from silence to singing aloud. Shouldn't "*A*" already be there? But note that if "*A*" must be present in one's being willing to sing "*A*" and singing it, it must also be present in being willing to sing "*A*" and not singing it. Similarly, in wishing or believing something to be a fact, we want the fact to be there as a sort of shadow. Between being willing to sing "*A*" and singing it we want a shadowy transition, effected by singing to oneself, or by making up one's mind to sing it, or by remembering it. And between the question "Are you willing to sing '*A*'?" and the answer we also want an intermediary. Here *understanding* is the shadow. Seemingly what we always want is that in the willingness what one is willing to do is already done, and similarly for a wish and what is wished for.

"Preparing to do so-and-so" is precisely similar to "being willing," "wishing," etc. About this expression the same puzzle can be produced as in the case of the latter. My method is to take a parallel case where one is not initially puzzled and get the same puzzle about it as in cases where one is always puzzled. Preparing to do so-and-so and what is to be done are very different. We are all willing to admit this. Yet we look for the thing prepared for in the preparation. But if the preparation is something different from what is prepared for, what has it to do with it? What is the relation between preparation and the thing prepared for? Must we know what the thing prepared for is by looking at the preparation?

It is not a hypothesis that we are preparing for just this. We do not say we *believe* we are preparing to do this. If "I prepare to do this" meant only that I do something which past experience has shown it likely to be helpful, then our puzzle would disappear. But we do not call an action a preparation for another action merely because experience has shown the one would be useful for the other. What is wanted is something more. One way out of the difficulty is this: to call "preparation to sing '*A*' ", for example, writing down the score. Then there cannot be such a question as "How does one know one is preparing for it?" For preparing is writing down the score. If one knows what is now done and says "This is a preparation", the question does not arise. The case where there is a definition of what preparation consists in is a very much simplified one. The definition shows what answer one could get to the question "How do you know you are preparing for this?"

The puzzle does not arise in the case of a man preparing a bit to put on the lathe, but only where we contemplate the expression "preparation for this." Then we ask "How does he know he is preparing for this?" And if we can ask this question we can also ask "*Does* he know?" The difficulty which makes one ask this is the same: that whatever he does in preparation is different from what he is preparing to do. What we would call the criterion for his knowing what he is preparing to sing, for example, might be his explanation of the words he uses—*not* doing it, but giving an explanation. And this is a grammatical explanation. The answer given when he is asked to explain shows what the question "Does he know what he is preparing to do?" means. Similarly, the criterion for knowing what one expects, e.g., for knowing that I expect Skinner to come into the room, is an explanation of what is meant by "coming into the room" and pointing to Skinner outside. The fallacy we are tempted to commit is thinking that we do not know what we expect unless what is expected already has happened, or that we do not know what we are preparing to do unless what is prepared for is already done. The difficulty which stands in our way is that in our ordinary language a phrase, say, "preparing for," is used both in the case where the relevant action is done and in the case where it is not done, e.g., when one prepares to sing "*A*" and sings it, and when one prepares to sing "*A*" and does not sing it.

This is the whole difficulty, and we can see it best in the parallel case of negation: Not (this table is green). How is it that I express a fact about this table if I say it is not green? At first it appears that the sentence "This table is green" which is prefaced by negation cannot have any meaning since it refers to what does not exist. Some people have said "not green" means the same as "brown or red or blue or . . ." because they wished to avoid negation. But one can explain what "table" and "green" mean; and that is enough. It might be objected that this explanation is inadequate, that though it gets nearer the correct explanation it does not quite suffice. The difficulty is that we do not see how the word "not" is used: the assertion "This table is green" is not part of the assertion "This table is not green." To throw light on this remark, let us treat of negation in a picture language instead of in a word language. That is, let us draw a picture instead of using the word "not". How are you to obey me and how are you to know what not to do when in this language I say "In fencing don't take this position"? Suppose I proceed as follows in teaching you gymnastics: you copy me when I take up a certain position, and I assume a different kind of position to indicate that you can do what you please but not this. What I have done in this language to describe what is to be done is part of the symbolism. In this language "*p*" comes in in two ways, as asserted, and in "not-*p*"; in the one case with "Do this" and in the other case with "Don't do this." Now where does the "this" appear? Only in what I actually do. It is part of the language, and that part which is a bodily sign. The connection between thinking what is not the case and that case is in the *sign*.

10 One of the questions which comes up in discussion of memory is the following: Can I and my body be identified? It might be argued that "I" is not used to refer to my body because two persons might have the same body. I want to say that realizing that the word "I" does not mean the same as "my body," i.e., that it is used differently, does not mean that a new entity besides the body, the ego, has been discovered. The argument that since *I* cannot be identified with *my body,* there must be something else, gives the impression of reporting a discovery. All that has been discovered is that "I" is not used in the same way as "my body". If I were to say (as I would not) that my body has toothache, instead of "I have toothache", this would merely express something wrongly. It would imply that there is no such thing as *I*, and would come to replacing "I" by "my body." This is like the mathematician's saying (rightly) that there is no such thing as number as an entity, and then saying wrongly that numbers are scratches on paper.

Consider the idea that the ego is a sort of collection of memories. Suppose a man had a peculiar memory in that when asked "What did you do yesterday?" he gave different replies on alternate days, on one day a description of day before yesterday's events, and on the next day a description of the events on the day after, alternating regularly in his replies. One might say there were two people. (Of course it is a question of terminology whether one says Dr. Jekyll and Mr. Hyde are two persons or one.) Suppose that suddenly both of them showed the ordinary phenomenon of memory. Here we might feel tempted to say that one of them had left. Compare this situation with that of a person with normal memory who did not remember a thing about yesterday, either because he had slept all day or had had no occasion to remember. One would not feel inclined to say he had died that day. In both instances the more one behaves *like* the way one did on the other days, the more one would be inclined to say that it was the *same* person. A different criterion for identifying a person might be used in the imaginable case of a race of people who looked pretty much like each other except for the color of the hair, tone of voice, and the fact that each had a different number of crosses on his forehead. Suppose that each person had a definite character with specific traits such as slowness, etc., and that these characters shifted about from body to body. We should be tempted to give the name of the person to the character rather than to the body which had it. In such a world where groups of traits traveled from body to body, any confusion over one body uttering another body's usual answers would be dispelled by asking what the person's memories were, and thereby distinguishing between people by means of their memories.

Suppose one replied to the question, "Who remembers last year's earthquake?" with "I", pointing to the body. (Note that there is no hypothesis about this answer, otherwise one could say "I think it is I who remembers".) There is a queer mistake, hard to explain, in considering that pointing to the body when one answers "I" is an indirect way of pointing to the self. It is connected with counting objects in visual space, in which we understand what we oppose our bodies to. We can count bodies, but how do we count selves? What do I oppose my self to? Since other people's names refer to bodies, what is the use of names for selves? We are inclined to say names for selves refer to hypothetical entities connected with bodies. But this is a mistake. To suppose that each of you has a self like myself is like supposing everybody has a shilling though I don't know they have. I only know that I myself have a shilling. Roughly speaking, the supposition of my having a shilling is a picture; the act of supposing might be done by a drawing. But what is the supposition like that each of you has a self like myself? When one talks of selves of other people, one thinks of some sort of spatial relation. Let us examine the supposition that each of us has a shilling to see how it differs from the supposition that each of us has a self. Do you see that it is essential to the first supposition that we be able to draw a shilling? Part of the game of supposing that other people have a shilling is being able to make a picture. The sense of the word "shilling" is given by the use we make of the word, and part of what we do with any sentence containing the word would be showing a picture. How do we use a sentence? A sentence has not got its sense "behind" it; it has it in the calculus in which it is used. The sentence, "Each of you has a self", sounded originally like the sentence, "Each of you has a shilling". But when you see how different the sentences are, "Each of you has a self" immediately loses some of its interest. That the supposition of having a self is very different from that of having a shilling of course does not mean that the supposition that other people have selves is necessarily nonsense. It might mean that other people are alive.

Suppose now that I changed my body in the course of a dream, and that the new body replied to the question, "Who had the dream?" with "I had it". There would be no question whether the new body had had it, and no question as to *who* had had it. Next suppose I say "Although I cannot imagine other people without their bodies, I could nevertheless imagine myself without my body". It might seem as though there was a sort of knowing expressible by saying "I know who had the dream, and where he is, namely, in this body". But is it sense to say "If I did not have a body I would still know that it was I who had the dream"? What would it be like to *know* I had a dream without having a body? If selves had no bodies, how should we make ourselves understood? Of course we could imagine that voices came from various places. But what use would the word "I" have, inasmuch as the same voice might be heard in several places? The fact that it makes sense to suppose that I change my body, but that it does not make sense to suppose that I have a self without a body, shows that the word "I" cannot be replaced by "this body"; and at the same time it shows that "I" only has meaning with reference to a body. A parallel in chess is that although the king is not to be identified with this piece of wood, at the same time one cannot talk of a pure king of chess which has no mark or symbol corresponding to it. The use of the word "I" depends on an experienced correlation between the mouth and certain other parts of the body. This is clear in the case where the criterion of a person's having pain when *his* hand is pinched is that the words come out of his mouth. It was in order to show that "I" would have no meaning without such a correlation that I tried to describe a case where there seemed to be a use of the word "I" [viz., imagining oneself without a body] but where closer investigation indicated there was not. Since "I" and "this body", like "the king of chess" and "the wooden piece", cannot be interchanged, it is incorrect to say that pointing to this body is an indirect way of pointing to me. Pointing to this body and to me are again different.

11 When I gave the example of having a toothache in someone else's tooth, it was to show that under certain circumstances one might be strongly tempted to do away with the simple use of "I". My idea was

to show that our use of this word is suggested by certain invariable experiences, and that if we imagine these experiences changed, the ordinary use of the word "I" breaks up and we see that the "I" notation is not the only notation which can be used.

Now it is a confusion to persist in the idea that in omitting something from our language we have thereby mangled the other language, i.e., that certain changes in our symbolism are really omissions. Thus we feel that if "I" were left out, the language which remains would be incomplete. We think we describe phenomena incompletely if we leave out personal pronouns, as though we would thus omit pointing to something, the personality, which "I" in our present language points to. But this is not so. One symbolism is just as good as the next. The word "I" is one symbol among others having a *practical* use, and could be discarded when not necessary for practical speech. It does not stand out among all other words we use in practical life unless we begin using it as Descartes did. I have tried to convince you of just the opposite of Descartes' emphasis on "I".

Whenever we feel that our language is inadequate to describe a situation, at bottom there will be a misunderstanding of a simple sort. One often has the experience of trying to give an account of what one actually sees in looking about one, say, the changing sky, and of feeling that there aren't enough words to describe it. One then tends to become fundamentally dissatisfied with language. We are comparing the case with something it cannot be compared with. It is like saying of falling raindrops, "Our vision is so inadequate that we cannot say how many raindrops we saw, though surely we did see a specific number". The fact is that it makes no sense to talk of the number of drops we saw. There is similar nonsense in saying "It passed too quickly for me to see. It might have gone more slowly." But too quickly for what? Surely it did not go too quickly for you to see what you did see. What could be meant by "*It* might have gone more slowly"?

12 It is queer that we should say what it is that is impossible, e.g., that the mantel piece cannot be yellow and green at the same time. In speaking of that which is impossible it seems as though we are conceiving the inconceivable. When we say a thing cannot be green and yellow at the same time we are excluding something, but what? Were we to find something which we described as green and yellow we would immediately say this was not an excluded case. We have not excluded any case at all, but rather the use of an expression. And what

we exclude has no semblance of sense. Most of us think that there is nonsense which makes sense and nonsense which does not—that it is nonsense in a different way to say "This is green and yellow at the same time" from saying "Ab sur ah". But these are nonsense in the same sense, the only difference being in the jingle of the words.

Rules for the use of words can exclude certain combinations, and this in two ways: (1) when what is excluded is recognized as nonsense as soon as it is heard, (2) where operations are required to enable us to recognize it as nonsense. The fact that negation of a complex tautology is a contradiction is discovered by the same means as $x^2 + 6x + 7$ *has two integral roots* is found to be true—by means of operations. We might think, for example, that "S has x pairs of shoes, where $x^2 = 2$" made sense because we would have a feeling that we could get sense out of it by solving the quadratic equation $x^2 = 2$. The fact that we do not see what the result is, is one reason for thinking we could call the sentence a different sort of nonsense from "tables, chairs, shoes". The word "nonsense" is used to exclude certain things, and for different reasons. But it cannot be the case that an expression is excluded and yet not quite excluded—excluded because it stands for the impossible, and not quite excluded because in excluding it we have to think the impossible. We exclude such sentences as "It is both green and yellow" because we do not want to use them. Of course we could give these sentences sense. I said earlier that what is possible or impossible is an arbitrary matter. We could make it a rule, for example, that "green and yellow can be in the same place at the same time" is to make sense.

13 We tend to think of a possibility as something in nature, something we are able to imagine. Roughly, when one talks of possibility, one is making use of a picture. When we say "This is possible", what is actual is a certain picture. Suppose that when I say it is possible for me to sing "God Save the King" I mean that I visualize the score. The score is the picture made use of. But someone will ask, "What is it a picture of? Of that which doesn't exist?" There is a difficulty which it is important to see: How is it that when we say a certain thing is possible we may know that what is possible is not what is actually the case and yet know what the possibility is a shadow of? We are tempted to say of the possibility that it is potentially present. The sentence "It is potentially present" makes it seem as though we had given an explanation over and above our saying it is possible for us to

do a certain thing. But actually we have merely replaced one expression by another. Similarly, if the word "not" is explained by saying that not-*p* is true when *p* is false, all that has been done is to replace "not-*p* is true" by a different phrase. We can make a word clear by means of another word only insofar as we make it clear by the grammar of the other word. The words "not" and "negation" are so related that we could replace the use of the one by the use of the other. The mistake we are liable to make is to think that the one word "negation", say, describes a phenomenon to which the grammar of the other word is to conform. But the grammar of the one word must conform to the *grammar* of the other, not to a phenomenon. We have the idea that we are putting up a standard of usage in nature, but in fact we are only putting up a standard of usage in grammar.

14 We cannot say of a grammatical rule that it conforms to or contradicts a fact. The rules of grammar are independent of the facts we describe in our language. To say that a grammatical rule is independent of facts is merely to remind us of something we might forget. And the point of remarking it is to warn us against a peculiar misunderstanding.

By means of an example where no confusion is likely to arise, viz., the length of a rod which serves as a unit of length, we can explain in what way the length is arbitrary and in what way it is not arbitrary. In the sense that one chooses this length and not that because of practical considerations, the length is obviously not arbitrary. The point here of saying it is arbitrary is to combat one particular misunderstanding, [namely that a convention can agree or fail to agree with facts]. Pliny said that after the number 10 the numbers repeat themselves. He thought that they did this because of the way they are written down, this latter being determined by the numerical facts. This is a mistake, for the numeral system is arbitrary. On Pliny's view a different notation would disagree with numerical facts because it differs from his system which supposedly agrees with them. The same error Pliny made about numbers could be translated into an error about lengths, viz., that after a certain point, say 12 inches, lengths repeat.

Some sentences are propositions, and other sentences look like propositions and are not. Whether they are or not depends on conventions. What are the conventions determining that a sentence is a proposition? Sentences which themselves state conventions seem not to be propositions. Nevertheless we tend to think they must conform to certain facts, in which case they would not be arbitrary. Consider a statement about primary colors. Suppose that whatever is called a color is one of these six or a mixture of them. This means that in our grammar of colors it makes no sense to talk of a seventh primary color since we have only six primary-color words. The expression "primary color number seven" has no meaning. Some people would say this means that the grammar of "color" must conform to certain facts of nature. But there is no parallelism between "There is no seventh primary color" and "There is no 6′2″ man who can be fitted with the six sizes of suits manufactured". It might well be asked, "Why not have a seventh primary color if the grammar of 'color' is arbitrary?" The reply would be another question: Will the new scheme come into any conflict with observed laws? How could it? It is not a fact of nature that seven primary colors cannot be arranged on the corners of a regular polyhedron. What it would be reasonable to ask is whether there would be any use for "seventh primary color".

Sentences which by present conventions clearly make no sense, e.g., that a man travels around the earth along the following route,

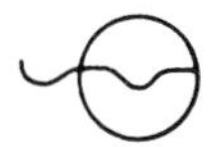

can of course be given a sense. It is because whatever is said can be given sense that the conventions adopted are called arbitrary. It might be objected that though it *can* be given sense, one part of the grammar must be analogous to other parts. Let us examine this demand. Suppose someone said that since our space has three dimensions we can therefore describe the path of a particle by the increases and decreases of three coordinates but not of four. He will claim that this imposes a restriction on our grammar since it is of the nature of space to have three dimensions. I would reply, "Aren't four dimensions just as good? The fourth variable could be darkness and light. If the particle gets darker, the fourth variable has a lower value." Another example of a description being given sense is the following: S says science and religion are coming to agree more and more, as the fourth dimension makes it easy to understand how Christ came into a room without passing through the door. He came by the fourth dimension. S thinks this explanation makes the statement easier to understand. And of course it *could* be described in these terms. Suppose we take time as our fourth dimension, measuring the point where Christ is by a watch. Where He is spatially could be described by three dimensions and where he is between vanishing and reappearing by the fourth. Now someone might object that he wants the new grammar to be analogous to the old. So let's have analogy so far as the formula is concerned, by giving the distance as $\sqrt{x_1^2+x_2^2+x_3^2+x_4^2}$. This of course is not the only thing that could be called analogous to distance with three dimensions. Usually there is one analogy that psychologically appeals to us most. We can do anything we please, but we'll see that certain conventions are too cumbrous to be used. Whether they are cumbrous or not depends upon our nature and the natural facts, e.g., that bodies do not vanish in one place and appear in another. If this appeared to happen we could say our vision failed us, but we *need not* say this.

15 To return to discussion of the general notion of a proposition. Have we got any general notion? What would we do if we had to explain what a proposition is? Propositions do not all have something in common, but are a family of things having overlapping likenesses. We can make sub-groups of this family, e.g., hypotheses such as "There is a window over there", and, by contrast, descriptions of immediate experience such as "I see a light patch in a dark surrounding", which form another group. What do I mean by a description of immediate experience? I have given examples, but in what sense have I got a general notion which would draw a line around this group? When asked to make groupings within this family, I gave and could give, only examples. Two questions arise: Have I the right to talk of this group without giving a general explanation? In what sense have I a general idea of this group apart from examples given? A general idea which is not a general symbol is of no use. That is, any general idea I claim to have of a proposition describing immediate experience is of no use unless used as a symbol. If it is taken as a picture or a hypothetical brain state and not used in the symbolism it is of no interest. My procedure is to look at the use made of the idea, *proposition describing immediate experience*. The use of the idea was explained by examples. These examples are not a clumsy way of describing it. I do an analogous thing in answer to the question "What is your idea of a formula?" I describe what I mean by a formula by writing down formulae. The word "formula", and formulae, are given, and nothing else is exhibited. When you say you understand the word "formula" I can charge you with not understanding me if what you do contradicts my explanation.

The word "proposition" is explained in the way "game" and "sense" are, by grouping examples. The examples give a clear enough idea. A person who has drawn the line by a definition might be held to have a clearer idea. And if you like you can give a definition; but as a usual thing one does not.

It has *seemed* that although we could not say what a proposition is in terms of a general definition, we could use a general idea, which we had, in a calculus. Frege and Russell made up a calculus which looked to be *the* calculus underlying the correct use of language. Logicians seem to have given a clear-cut definition, whereas I have explained the idea of a proposition only by giving examples. How does what I have done compare with the clear-cut idea which logicians talk about? The logical calculus is clear-cut enough, but it is not fundamental and it is not very applicable. For a definition would apply to some things fairly well and to others less and less well.

16 The notation "$(x)fx$" of Russell's calculus, meaning "for all things so-and-so is the case", needs to be examined in each instance for sense. It was originally taken to symbolize statements of ordinary language such as "All men are mortal" and "All men here wear grey flannels". Then it was extended to talk of "all numbers in the series of cardinals", "all points of this surface", where very different grammars apply. The grammars of generality, and of negation, are ambiguous in an incredible way. Suppose I translate "This square is white" as "All points of this square are white". According to Russell, this would come to the same as saying "There isn't a thing which is a point in this square and is not white". And "This square is not entirely white" would come to "There is at least one point which is nonwhite". Now what is it *like* for *one point* to be nonwhite? We might *give* it a sense. But when we translate "This square is white" into "All points are white" we cannot without further conventions give sense to "One point is nonwhite". To see the difference among grammars, ask yourself how you would verify such propositions as "All points are white" and "In all the circles in the square there is a black point in the middle". In the case of both "$(x)fx$" and "$(\exists x)fx$" Russell takes the "x" inside the bracket to stand for a *thing*. What sense does it make to say "There is a *thing* which is a black point and which is in the square"? Russell says "I met a man" = "I met a thing which is a man", and "All men wear grey flannels" means "All things that are men wear grey flannels", or alternatively, "There is no

thing which is a man and does not wear grey flannels''. Can one talk of a *thing* which is a man? And are we to go through all the things to determine that there is no thing which is a man and does not wear grey flannels? The ''*x*'' inside the bracket stands for men, not things.

Like ''all'', negation also has different grammars. It has been asked whether the negation of a proposition comes to the same as a disjunction of propositions. In certain cases it does, e.g., ''This has one of the primary colors, but not red'', which means ''This is white or yellow or green or blue or black''. But there is *no* disjunction corresponding to ''Smith is not in this room''. The ''and so on'' of the supposed translation into ''Smith is there or there or . . . and so on'' is not a disjunction.

17 It is in a sense arbitrary what is called possible and what is not called possible. We say that though no man sits in this chair somebody could. This means roughly, ''The sentence 'Somebody sits in this chair' makes sense'', that is, there is a logical possibility of someone sitting in it. It is theoretically possible, i.e., possible in a theory, for some hydrogen to have six times the normal valency. This might be possible in certain theories but impractical. Some theories are practical and some impractical. Impractical systems are rejected, and the rejection is treated as though what is rejected is something false. Rejection of a grammatical system is like the rejection of a standard of length, [and acceptance of a grammar, a symbolism, is like acceptance of a standard of length.] Consider another comparison: that every proposition is a picture of reality. Here, in the comparison with pictures, we have an extension of the use of the word ''picture'' that we are very much inclined to accept. Such extensions may be very valuable in showing transitions between examples, for the examples form a family looking different at the outskirts. What the family looks like, e.g., the family of plants, will depend on what we take as a standard.

The fallacy we want to avoid is this: when we reject some form of symbolism, we are inclined to look at it as though we had rejected a proposition as false. It is wrong to treat the rejection of a unit of measure as though it were rejection of the proposition ''The chair is three feet high rather than two''. This confusion pervades all of philosophy. It is the same confusion that considers a philosophical problem as though such a problem concerned a fact of the world instead of a matter of expression.

I have indicated that there is a difference between rejecting a hypothesis as false and rejecting a symbolism as impractical. But there are transitions from one to the other. Suppose that a planet which according to a certain hypothesis describes an ellipse does in fact not do so. We should then say that there must be another planet, unseen, acting on it. It is arbitrary whether we say our laws of orbit are right, that we merely do not see the planet acting on it, or that they are wrong. Here we have a transition between a hypothesis and a grammatical rule. If we say that whatever observations we make there is a planet nearby, we are laying this down as a rule of grammar; it describes no experience. We may then be forced to make a queer alteration. We would have to model everything else to account for it. (Consider the changes required by accepting the hypothesis that there is a hippopotamus in the room.) [To illustrate the different roles of proposition and rule of grammar], suppose the standard of a foot length was a rod in my room, and suppose the Greenwich rod agreed exactly with this rod. To say "The Greenwich rod is as a matter of fact a foot long" is to assert a proposition, whereas at present it does not make sense to say this. It is a definition.

We can draw the distinction between hypothesis and grammatical rule by means of the words "true" and "false" on the one hand, and "practical" and "impractical" on the other. We do not speak of propositions as being practical or impractical. The words "practical" and "impractical" characterize rules. A rule is not true or false. But now with hypotheses we use both pairs of words. One person says a hypothesis is wrong (when he is unwilling to remodel other things), another that it is impractical (acknowledging that he could remodel other things). Deciding whether a sentence is used as a hypothesis or as a grammatical rule is like deciding whether a game is chess, or a variety of chess distinguished by a new rule entering at a certain stage of the game. Until we get to that stage, there is no way of telling which game is being played by looking at the game.

Suppose someone had learned mechanics in such a way that all calculations were done with the three laws of motion and d'Alembert's law. Suppose he transformed these laws and found the law of transformation of energy. Here we may ask: Has he discovered a new bit of mechanics or a new bit of mathematics? Given a description of phenomena in accordance with Newtonian laws, another description will not be called a discovery in mechanics if it is impractical. But it might be a new *mathematics*. He has made a new game.

There are cases where we would decline to call something a new bit of mechanics. Consider Hertz's mechanics, which substitutes for the three laws of Newton a single law: a form of the law of inertia, viz., that a system of material points either is in a state of rest or moves with a uniform velocity along a straightest line (the latter being already defined). Suppose now that someone built up a mechanics from the three laws. We might say, "This is no new mechanics—it is built on Hertz's". But it is a new bit of mathematics. A new bit of mathematics and a new bit of mechanics are not to be confused. To confuse them is to treat mathematics as it is treated in present foundations of mathematics—as though it can be reduced and reduced, and not as though it were something new. We cannot reduce mathematics; we can only make a new one. The size of a proof can be reduced, but not the body of mathematics. The same point can be made about chess. Suppose chess is defined by the way we move pieces, and that a new way of producing a certain move is discovered. This is not to reduce the old game; it is to make a new game.

To show what we do in philosophy, I compare playing a game with rules and just playing about, or playing in a way that is a transition between the two. What we are looking at is the use of language as compared to a game played according to rules. It is useful to exhibit the two extreme cases, the use of a sentence as a hypothesis and as a grammatical rule.

18 The laws of logic, e.g., excluded middle and contradiction, are arbitrary. This statement is a bit repulsive but nevertheless true. In discussing the foundations of mathematics the fact that these laws are arbitrary is important, for in mathematics contradiction is a bugbear. A contradiction is a proposition of the form *p and not-p*. To forbid its occurrence is to adopt one system of expression, which may recommend itself highly. This does not mean that we cannot use a contradiction. In fact it is used, for example, in the statement "I like it and don't like it". To the objection that "contradiction" is not used to apply to such a case, that "I do" is not contradicted by "I don't", I admit this is true if we take our system as primary. If we say a thing *can't* at the same time be both red and not-red, we mean that in *our* system we have not given this any meaning. An adopted system of expression is like an adopted measuring rod. Now in describing the application of a rod in certain cases we leave open the way it is to be used in analogous cases. We might mean by the real length of a body that length which a measuring rod made of iron will show when it is the same

temperature as the body. Or we *could* call the real length all the different readings at various temperatures. The objection that "I like it and don't like it" is not a case to which "contradiction" applies is paralleled by the objection that a measuring rod is useless unless it is rigid. But in some cases we might *want* elasticity. And a contradictory order, of which it might be said that it is not an order at all, could be used to produce uncertainty. Why not have a mathematics full of contradictions? We do often use a double negation, $\sim\sim p$, to mean $\sim p$. A person who says we do not mean it in this way is saying there are different kinds of double negation. This is to treat double negation as a fact of nature which in one case gives negation and in another case does not. The law of contradiction can, but need not, be used as a law of our expression. Contradiction can be dealt with in mathematics either as something forbidden or as something allowed. $2+2=4$ and $2+2=5$ together might be useless but not false. Together they would give a new mathematics.

I want to comment further on the rejection of "I like it and don't like it" as being a contradiction, the objection being that if we do use it we cannot do so in the same way as we use an ordinary noncontradiction. Now what *can't* you do? And what is the obstacle which prevents you? The word "application" brings out your objection. You say, "We can make a system of signs using '*p and* not-*p*', but the application will be different". But how can you talk of a system of signs without talking of the application, as if the application of the law of contradiction were independent of the law? Suppose I said my hands cover each other completely when I put them together in one way but not in the other. Then suppose I said they cover each other in both cases. Someone would object that "cover" is in the last case used in a different way. I would ask, "How do you wish to explain 'covering', with or without reference to something that covers?" In saying that two hands cannot cover each other in the same way if superimposed differently, has "covering" been defined independently?

The same objection and the same questions arise in the following circumstance. You object that the arithmetic in which $2+2=5$ cannot be applied in the same way as that in which $2+2=4$ can be. Has "the same way" been defined? Again, suppose I said "If a man is as short as I am he can pass through the doorway, but if he is 8 feet tall he could also pass through it but not in the same way". If I can sensibly use the phrase "in the same way", that way must have been given me independently of the notion of passing through the doorway. To illustrate the point with another example: suppose I project the length *A* on the line below

A

b b

and claim that I could also have projected it on *b-b*. Someone objects that I cannot do this in the same way. But what is the same way? He has left a loophole in his *expression.* If my description of the projection did not describe the whole figure, i.e., if the way in which I projected *A* did not reach to the projected line, then I should have the line projected, the projection, and the *way* of projection. If, on the other hand, in the word "way" I assumed the whole figure, then it could not with sense be said that I could not have projected it on *b-b* in the same way. The phrase "the way" has to apply to both in order for it to make sense to say that here this way could not be taken.

To return to the objection that "I like it and don't like it" is not used in the same way as a contradiction. What is it that we call the law of contradiction?—the formula, or the formula plus the application? If we mean the latter, then we cannot talk of the application as though it were independent of the law. The word "way" corresponds to the word "analogous", which means "something else", not the same thing over again. We cannot wish to explain "a way" independently if the way is *included in* what is being described. If we say we cannot apply logic in which contradictions are allowed *in the same way,* what makes this misleading is that there seems to be an obstacle. A word is being used for leaving something open and at the same time for closing it. What seems to be assumed is that a way has been described and that one could not get to the end of it. In fact a way has not been described. One cannot reasonably object to not reaching the end of the way if in giving the way one gives also the end of the way. There is no sense in talking of a way if there is only one end and a different end is precluded. In using the word "different" you have provided a loophole and precluded it.

PART III

Wittgenstein's Lectures

1934–35

From the notes of Margaret Macdonald and Alice Ambrose

Michaelmas Term

1934

Lecture I

What we say will be easy, but to know why we say it will be very difficult.

I shall begin with the general idea of a proposition. First, have we such an idea at all? Most definitions of a proposition given in logic books, as what is true or false, or as the expression of a thought, are futile. For we do not understand the terms of the definition. If one wants to explain what a proposition is, one thing one might do is to give examples, by means of which it might be said one could get the general idea. Now what is the criterion for someone's having a general idea? For example, of a plant. If after being shown different plants, say a fern, a geranium, and many others, a person is asked to fetch a plant and he brings one that has not been shown him, say a violet, we say he has the general idea of *plant*. That is, we infer it from his behavior. But what is it we *infer?* We would be likely to answer that he has something in his mind, which we cannot perceive but he can. If I am to explain what it is to perceive *my* general idea when I have a thought of a plant, there seems to be a difficulty of introspection. I would be likely to say that the idea passes so quickly that I cannot catch it. What I do catch is something irrelevant to the general idea, namely, accompanying images. Or I might say the idea is unconscious, that I don't actually perceive the idea when I use the word "plant". The word is used semi-automatically, so that the idea is in the mind in roughly the way the alphabet is when we are not repeating it but which we know we could repeat if required. We tend to think of the mind as a sort of receptacle in which things are stored. To say it is in the mind in this sense is a hypothesis. "In the mind", "in the head" are phrases used to denote a model, like the model a scientist uses to explain electricity. A model is part of the symbolism in which a hypothesis is stated. A hypothesis, among other things, serves to express expectations (here, what the man who has the idea of *plant* in his mind will do when asked to bring a plant).

Another notion we have about a general idea is that it is a sort of general picture, or composite photograph, with unclear outlines. It is dangerous to be tempted to think this, for if you examine yourself when you have, say, the general idea of a plant, you will usually find you have no such inward visual experience, either a composite image or the image of a particular plant or leaf. Realizing this may make you say that there must be something like it, something less crude. This is a sign of a philosophical problem. The account of what happens in your mind when you hear or use a sentence in which the word "plant" occurs, provided you understand the sentence and do not utter it like a parrot, will very likely be the following: (1) there is some image in the mind, (2) on examination no image is found, (3) the general idea, if not an image, must be something more subtle. The queer thing is that this subtle thing is never found. There may be nothing at all going on in your mind, and still you may not talk like a parrot.

What makes us believe that there must be such a general idea in our minds? First of all, because it is not enough for the general idea of *plant* that we should be acquainted with particular plants, say violets and roses. And we might have the idea of these without having the general idea of a plant at all. We think there must be something going on in one's mind for one to *understand* the word "plant". We are inclined to say that what we mean by one's understanding the word is a process in the mind. But *when* does this activity take place? Is it when the words are being heard, or immediately afterwards, or when? Is there something in the mind like a set of bells, so that when a word is heard one chimes, and when a sentence is heard several chime, one after the other? No, this is not the case. There is a way out of the difficulty of explaining what understanding is if we take "understanding a word" to mean, roughly, being able to use it. The point of this explanation is to replace "understanding a word" by "being able to use a word", which is not so easily thought of as denoting an activity. If we compare the questions "When do you understand a word?" and "When are you able to use a word?", we tend to answer the first by "while, or after use", but not the second. Being able to use the word is not an *accompaniment* of the use, as understanding seems to be.

Actually, words in sentences are very often accompanied by something or other, images or what not. For example, when asked what happens when you understand the word "red" you tend to answer "a red image", though you may get the image of something green. You do often have images, and this is one thing responsible for the notion that a general idea is in your mind, and that understanding a word is an activity accompanying it. Also, since talking can go on without thinking, as when we say something while thinking of something else, or speak without understanding, it is natural to think of speaking and understanding as two activities going on at the same time. But does it follow that understanding is an accompaniment of speaking? Suppose the order were given to say the alphabet, and that the alphabet was repeated in response to the order. In what way are the twenty-six letters involved in what was asked of the person and in his understanding of the order? That they are involved is clear. But does he have to repeat the alphabet inwardly in understanding the order? No. If there were someone who knew the alphabet as we know it except that he left out *z*, or who understood by "the alphabet" a different ordering of the twenty-six letters than the usual one, we might ask whether he understood by "the alphabet" the same thing as we do, or a different thing. If a different thing, then this difference should come into the understanding.

The word "understanding" is used in two different ways, one which seems to allude to processes accompanying hearing or uttering a word or sentence, and the other which appears to have nothing to do with an accompaniment but is something like being able to use the word or sentence. The expression "being able to" is such that our criteria for its use are doing the thing, having done it, saying we can, etc. It may be the case that whenever one hears a word a peculiar mental event occurs. I do not wish to deny this. Perhaps when you hear a word a light flashes inside your brain or something like a bell sounds. But *is* this experience what is meant by "understanding a word"? It is not, though something mental may be involved in understanding even though one cannot say precisely what. I am not trying to give a definition of the term "understanding". I am merely objecting to the idea that there *must* always be an experience there when we understand. For in great many cases in which we use the word "understand" we can substitute for it "knowing the use of". In others we cannot, for we use the word in a mixed sort of way. We are tempted, however, to say it is always this or always that. What gives one the right to say that in *all* cases so-and-so happens, unless it is experience? But it is not experience here that makes us say there must be a mental accompaniment when we understand a word.

Lecture II

When a child is able to use a word we say he has got hold of an idea, that he *understands* the word. This may be a way of saying that his being able to use the word is a hypothesis, and we then give him tests to find out whether we can rely on his using it aright in the future. The hypothesis that he has the general idea corresponds to the assumption of a hypothetical mechanism which we do not know because it is seen from the outside, like the works of a watch which no one has ever been able to open. The hands are seen to go round and we then make the hypothesis that they do so because there are works inside. And we could make a model of them. We very often think of ideas in this way. An idea is like a mechanism whose workings we do not know. Something mental does enter into understanding, but it does not enter in the way one would have expected, and it is not revealed to introspection.

Consider the statement of someone who says he knows what a plant is. When he claims this, was he referring to anything in his mind? His answer is No, but that when he considered whether he did know it or not, there was something going on in his mind. I wish to make the point that the definition was not in his mind consciously when he said he knew what a plant is, and to assert it is unconscious is like assuming a hypothetical mechanism. We look at the notion of having a general idea in two ways which are contradictory, (1) as a process happening roughly before, during, or directly after the word is used, and (2) as something like being able to use the word. We hardly ever use the word "understand" for a process happening while the word is uttered. In most cases it is used to mean being able to do so-and-so. When a man understands an order it is true that often certain pictures are present to his mind, though often not. If by "understand" is meant that such a picture is present, then the word is being used by me as it is practically never used.

Lecture III

The following questions are important: In what way does giving a definition show that one has a general idea? Does giving it prove that certain things *must* happen when the word is uttered? Or does having the general idea mean being able to give the definition? What is "being able"? Is it a disposition or is it a feeling accompanying the word? Is giving the definition the same as having the general idea? It has been held that they are different. Now is the difference that between giving the definition and being able to give it? If you ask me, "Are you able to lift this?", and I answer Yes, what does "being able to" mean? Suppose I try and do not succeed. *Was* I able to? There are two possible answers: (1) No, I was wrong, (2) Yes, I was able to, but now I cannot. If I answer that I was wrong, then the claim to be able to lift it was a hypothesis. The second answer is not a hypothesis.

In a case where you have already whistled to yourself, whistling to yourself is an accepted criterion of being able to whistle aloud. Here being able to whistle aloud is doing something slightly different. I call the latter a symptom of being able to do it. Giving a definition is a symptom of having a general idea.

Suppose you saw people playing with a ball and after seeing a hundred such games were asked to write down the rules of the game. It must be admitted that for ordinary games you could do this after a time. Now there are all sorts of intermediaries between playing games according to rules and just playing about. With our language it is similar; there will be cases where, after observing the use of a word, the rules are clear, and others where it is not. Consider the use of "able" in "able to lift", "able to eat a large dinner", "able to stick boring company". People say these all have something in common. But the similarities between their uses are instead *overlapping*.

Lecture IV

The phrase "being able to" in "being able to give a definition" means various things, (1) a conscious state of mind, (2) a state of the brain, shown, for example, by operating on the skull, (3) saying you could produce it if asked. "Having a general idea", as a single phrase, tempts us to think there is one phenomenon common to all cases of having a general idea; but the examination of particular cases shows otherwise. Consider two color patches, one red and one green. What these colors have in common is *being either red or green*. This is not to have something in common in the same sense as, say, two tables have something in common or as a centaur and a man have something in common. Most people think something like a feeling to be common to all cases of having a general idea. What is the difference between my having a general idea and my not having it? The difference is in my reactions.

The question, "Have we got a general idea of a plant?", is put wrongly. For it suggests that to answer it is a matter of taking a census. The difficulty is that when we began talking about general ideas we became uncertain what we meant by having a general idea. Can the question, "What is a general idea?", be answered? I will not answer it, but I will say something about the use of a word. In most cases where such a question is answered we usually do say something about usage. I will say this much: there are all sorts of cases in which one says "I have a general idea".

Lecture V

Let us return to the question whether we have a general idea of a proposition. The difficulty with giving an account of any general idea is that we try to combine two contradictory aspects of it, dynamic and static. In many cases we think of a thing in terms of a mechanism. The point of thinking of it in these terms is that from the way it looks and from certain tests I shall call static (handling, looking, etc.) and which are made before we test the behavior of the mechanism, we draw conclusions as to how it *will behave*. For example, when we have examined a screw and found, say, that its threads are not broken, we say we know what it is going to do if certain things are done to it. The way it looks or feels stands for the way it is going to behave. Yet it does not *follow* from the way it looks that it *will* behave so-and-so. For we can imagine a fountain pen, for example, which looks like mine and yet will not unscrew. We examine its screw and cap and predict that it will behave in such-and-such a way, but whether it *will* behave in this way is a hypothesis, a conjecture. There is *no* static test, i.e., one to which we submit a mechanism before it is put into use, which will enable us to *know* it is going to work in a certain way. This is always hypothetical. We may be wrong in expecting a certain behavior.

We are accustomed to think of things in terms of a very few definite possibilities. If two cylinders are such that one is smaller than the other, we say that one will turn inside the other. If it does not, we say something must be stopping it. It might be very puzzling why it does not turn, and we might say there must be a *cause* for its not turning. But what more does this mean than that in some circumstances it will turn and in others not. To say it will turn if nothing is wrong means nothing. Can't I assume that it does not turn? We do not have here a case of one thing following logically from another. It is a conjecture.

When we see a diagram of a wheel connected with a piston rod we have one idea of how it will behave. We do not assume that the wheel is made of dough or will suddenly become elliptical. Yet how do we know that these things will not happen? Suppose we reply that we *assume* the wheel remains *rigid*. What do we mean by its being rigid? Is it merely that it will behave in the assumed way, or is it something else? At first sight it seems as if there is one static test from which it follows that the behavior of the wheel and piston will be such-and-such. But now we find we are assuming rigidity, for which we have to give tests. There are static tests, such as trying to bend the thing. But from these tests it does not *follow* that it is rigid. It is a conjecture, but one which we always make; for we are accustomed to a certain mechanism corresponding to certain behavior. We cannot say that what we conjecture must happen if the body is rigid, since rigidity itself is something established by empirical tests. If our conclusions from tests were not conjectures, then "This is rigid" would have to contain the fact that the thing will do so-and-so. That is, what the thing is is the class of things it does.

Now for the application of these considerations to the notion of a general idea as a mechanism, which we tend to think of as static and at the same time conditioning what will happen. Our notion of it is of a mechanism from the existence of which it *follows* that we will use a general word in this way or that way. We have the wrong idea that the use of a word is like pulling a thread from a bobbin: it is all there and needs only to be unwound. Thus we talk about some uses of a word as being in accordance with the general idea and certain other uses not. If you ask me, "Is this a plant?", "Is that a plant?", and I answer Yes, or No, these answers show what my general idea is. But I do not have all these answers in my mind when I say I have an idea of a plant. Thus the general idea is looked on as something static, a disposition in the mind, to be tested by whether the answers to the question, "Is this a plant?", are in accordance with it.

What agrees or disagrees with the idea seems in some queer way to be contained in it, to follow from it. What agrees or disagrees seems to be compelled by the idea. I wish to show that we confuse two different things, a law of nature and a rule which we ourselves lay down. The statement that a mechanism made of iron will when tested in a certain way behave so-and-so is a law of nature. The statement that moving a line in the diagram of the mechanism alters the angle is a statement of geometry, not physics. The result of moving the line is in accordance with rules we have laid down. These rules are rules of our symbolism. Suppose that the lengths of bodies in this room were as a matter of fact multiples of the length of the arm. If we wanted to fix a unit for our measurement, it would be natural to choose the arm as the unit. But this is merely convenient; we are not *compelled* to do this. A philosopher would mix up the natural fact that bodies are multiples of the length of an arm with the fact that the arm is taken as the unit of measurement, which is a convention. They are utterly different, though intimately connected. One is a fact of experience, the other a rule of symbolism. The rule we lay down is the one most strongly suggested by the facts of experience. Geometry and arithmetic consist of nothing but rules of symbolism comparable to the rule which lays down the unit of length. Their relation to reality is that certain facts make certain geometries and arithmetics practical. If every time we counted 40 plus 20 we got 61, then our arithmetic would be awkward. We could make up an arithmetic in which this was true, and this is not to say that 61 is the same as 60. A rule is chosen because things have always been observed to behave in a certain way.

Suppose I gave you a sample, saying "This is *green*", and asked you to bring me something green. If you brought me something yellow and I said it did not agree with my idea of green, *am I describing a fact of nature*? No. To say that something yellow disagrees with the green sample is to give a rule about agreement. That yellow disagrees with green does not follow from anything in the nature of green or yellow. I could instead say that what disagrees with green is something that looks nasty with green, and yellow might be said to agree with green. If something is said to agree or disagree with an idea or thought, we do not *find* it agreeing or disagreeing. What are called agreement and disagreement is something laid down as a rule. And the rule is either useful or not. That a green or yellow agrees with the green sample is part of the geometry, not part of the dynamics, of *green;* that is, it is part of the grammar of "green", not a natural law.

Lecture VI

One of the chief difficulties we have with the notion of a general idea or with understanding a word is that we want it to be something present at some definite time, say when the word is understood, and the idea we have is supposed to have consequences and to act as time goes on. For example, the idea of a plant is supposed to enable me to identify something as a plant, bring a plant when ordered to, define "plant", etc.; and these phenomena are taken to agree or disagree with the idea. If by "general idea" we mean the *cause* of agreement and disagreement, there would be no difficulty, for then the idea would be an existent thing like an acid to which there is a reaction of some sort. But we do not want the relation between the idea and a phenomenon which agrees or disagrees with it to be a mere causal one. The agreement we want is not experiential at all. It is not a question of experience whether a thing will agree with our general idea, as it is with a mechanism about which we cannot predict with certainty. If we take the idea to be a natural phenomenon which can do such things as enable us to apply a general word or give a definition of it, our investigation of it is psychological. We are in the realm of hypotheses, about effects and causes, and not in the realm of the "must". But we are wrong to take the investigation of the general idea to be an investigation of the causes and effects of a natural phenomenon. We are mixing up two different things, a process which happens in our minds or brains, whose causes and effects can be studied by psychological methods as in other sciences, and certain rules which we lay down. To illustrate: Suppose someone is given the order, "Bring me something of this color", and I show him a blue sample. What agrees with it? There are different kinds of agreement, e.g., what matches, what contrasts pleasantly. The order might be obeyed by bringing any blue object, or by bringing something exactly similar, or by bringing something that looks well with the sample. We expect every idea to have tentacles or affinities, so that it predetermines what will satisfy it.

Why is it that the idea seems to be satisfied by some things and not by others? It is not a matter of experience that something satisfies it. In a way it must be satisfied beforehand. There is of course a sense in which experience enters. For example, if a piece of cloth is laid next to the sample and is seen to have the same color, we might say it is an experiment showing that the two are the same. But that this color is the same as the color of the sample is not shown by experiment. Whether it agrees or not is determined *a priori*. It is *a priori* that if you bring something blue, it will agree; this is not something you predict. Though this sounds like a prophecy, I know with certainty that the colors agree because I laid down a rule beforehand about what would be called agreement—about the use of the word "agreement". That they will agree is not known better after I juxtapose the two than before.

The question whether a man knows what he wishes is like the question whether one knows what will agree with a sample. When he does not know what he wishes, then what happens to satisfy his wish is a matter of experience. Here the fact that such-and-such satisfies his wish is not known beforehand; it is a hypothesis. Similarly, to say that the piece of cloth agrees with the sample is an experiential proposition, for it says nothing more than that it is blue. This I can know only by making a test. But to say that this color agrees with the color of the sample is a rule to the effect that this is what I call "blue". It is a rule about the use of a symbol. I could have made all sorts of rules, for example, that middle C on a piano agrees with blue. Then the blue patch is no longer a sample, but a word, or like a word. To say that it agrees with middle C would be a definition. If "agreement with an idea" does not mean a natural phenomenon, then propositions asserting such agreements are rules. And the rules do not *follow from* the idea. They are not got by analysis of the idea; *they constitute it.* They show the use of the word.

What idea do we have of the king of chess, and what is its relation to the rules of chess? The chess player has an idea of what the king will do. But what the king can do is laid down by the rules. Do these rules follow from the idea? Can I deduce the rules once I get hold of the idea in the chess player's mind? No. The rules are not something contained in the idea and got by analyzing it. They constitute it. I can give all the rules of chess in the form of a diagram illustrating the moves of the different pieces. Everything a piece does can be deduced from this, and an illegal move will disagree with this. The rules constitute the "freedom" of the pieces.

It seems at first sight that the rules for the use of a symbol are deducible from the idea connected with it. The idea always seems to be something containing its whole use, the use being something already there which we find by analysis. But the idea connected with the symbol is only another symbol. The rules are rules for the use of that symbol. The idea and the rules stand in the relation of a symbol and the rules for its use. So far as the idea is a static mechanism, what follows from it is hypothetical, and so far as it is not, what follows is *a priori*. We can say *a priori* only what we ourselves have laid down.

The following case seems to contradict the claim that the use of a word does not follow from the idea: by an example, i.e., by an ostensive definition, we are able to give a person an idea of *red,* say. We show him the *meaning* of the word "red". If we can give the meaning by ostensive definition, then the correct use will follow from its meaning and not from the rules. The correct use of the word "red" is thought of as a consequence of its meaning, which is given in one act, all at once. And this is inconsistent with my saying that the rules constitute the idea and do not follow from it. *However,* note that the use of the word is not actually fixed by giving someone, by ostensive definition, what is supposedly the meaning. For he may now use "red" when he sees a square.

Lecture VII

We have been puzzled by the notion that when we understand a word the idea we have makes us use the word in a particular way. It is as if the idea contains the use, which is then spread out in time. I tried to trace this notion to that of a mechanism. Now what a mechanism does does not follow from what it is in any important sense. What it will do can only be conjectured from what it is, unless we already include in what it is what it does. Before we realize that we can only *hypothetically* infer what a mechanism does from what it is, we tend to compare an idea with a mechanism. This notion must be discarded. Those propositions which seem to be analogous to experiential propositions about mechanisms, for example, that such-and-such an action or use will "agree with" or "follow from" an idea, are not really analogous, since they are *rules*. And this is why they have the appearance of certainty which the analogous propositions about mechanisms do not have. These then are not experiential propositions as to how the idea will behave when confronted by so-and-so, for this is conjecture. The *a priori* statements about agreement of something with an idea are misleadingly put in the form that the thing *will* agree with the idea, as though it were a question of time. But time does not enter in. One might as well say that 2 plus 2 made 4 yesterday or would a thousand years hence.

What we are apt to confuse is the idea as a state of mind occurring at a particular time and the use we make of that idea. The reason for the notion that the idea as static—as something before the mind's eye—has its uses contained in it and needs only to be spread out in time for its uses to be revealed is this: that in the case of many ideas there is one preeminent use. Consider a rudimentary case: a green image taken as the idea of the color green, and used as a sample. (A piece of paper could be used as a sample just as well as an image.) In a vast number of cases we are inclined to think that a particular idea can only be used in a particular way. The use of a sample green image or object is most often in comparing and copying it. This is the main use, but obviously not the only use. There are even many different kinds of comparing and copying. There is rough and exact copying, comparison of this green with other greens, comparison of two colors by means of a color wheel with respect to the amount of yellow they contain. Among the varieties of comparing and copying there will be some common or garden ones, such as the rough comparisons we make in ordinary life. An idea, if we mean something static, is a means of operating with language, and in all sorts of different ways, although as a matter of fact it is a means which is almost always used in one way. As soon as we see that this use is only one of lots of uses, we see that the idea plays the role of a symbol.

As another example, consider the idea of a circle. In general the criterion of whether one has this idea is being able to draw or copy a circle. By copying everybody means roughly the same thing. But projecting the circle so that it becomes an ellipse is also a way of copying. In general people mean one thing by a circle and another by an ellipse. But why shouldn't the idea of an ellipse be a circle having the diameter of the long axis? Nevertheless, that we could use a circle in a way analogous to an ellipse never occurs to us. Copying almost always means the same thing. One use is preeminent, just as the sign ⟶ is always used to indicate that one is to go forward, not backward. Even in the case of a savage tribe whose language we did not know, we should assume (rightly) that an arrow pointing forward always meant going forward. If I pointed in this way

it would be interpreted to mean "Forward, then to the right, then forward again", and not as an order to turn round or go backwards. Similarly, if I showed a person a red sample and asked him to bring me something red, he would do so. But he need not. He might bring me something having the complementary color, though this is not usual. One use is not more direct than another—only more usual. We are extraordinarily affected by the way in which we do in fact react to a sign. The result is that certain ideas stand to us for certain uses because that is how we usually apply them. We therefore think that those ideas have that most usual use *in* them, though they could perfectly well be imagined to have another use.

Let us consider the word "understand", about which questions arise similar to those about having a general idea. When a person is given an order and says he understands, suppose we say that what he means by understanding is something going on roughly at the time of the order. Having understood is said to make him able to carry out the order. Understanding seems to be a state of mind having certain actions as consequences. Certain actions will seem to be in accordance with it and certain others in discordance with it. Understanding, i.e., getting hold of an idea, can be looked at in two different ways, (1) as being a state of mind or mental process, (2) as the use made of the idea. Note that whatever the state of mind is, it does not necessarily agree or disagree with any action done. Suppose you said "Bring me a chair", and my understanding the order meant having an image of a chair. In such a case I usually do bring a chair—I don't go on to paint it or break it or do anything else. But must the action of bringing a chair agree with my visual image of a chair or even of myself bringing a chair? Cannot the two be compared in totally different ways? In taking understanding an order to be a state of mind which is had at the time of understanding it, it does not follow that you must make just that use of the idea which you generally do make. And if what you will do on understanding it is not to be merely a conjecture, you must lay down what will agree or disagree with your idea.

The straightforward answer to the question whether having an idea makes us do certain things, whether it contains in itself its own uses, is that it does not. To say why we asked such a question at all, or thought that there was a problem, I reminded you of our tendency to look on an idea as a mechanism, especially in cases where we are inclined to think of the one use the idea usually has.

The question has been raised as to what a convention is. It is one of two things, a rule or a training. A convention is established by saying something in words, for example, "Whenever I clap my hands once please go to the door, and if twice, please go away from the door". If by a convention is meant something laid down by a sign, does this mean that one *could* lay down another rule? By a convention I mean that the use of a sign is in accordance with language habits or training. There can be a chain of conventions at the bottom of which is a language habit or training to react in certain ways. These latter we do not usually call conventions, but rather only those which are given by signs. One can say these signs play the role they do because of certain habitual ways of acting.

Lecture VIII

It is felt to be a difficulty that a rule should be given in signs which do not themselves contain their use, so that a gap exists between a rule and its application. But this is not a problem but a mental cramp. That this is so appears on asking when this problem strikes one. It is never when we lay down the rule or apply the rule. We are only troubled when we look at a rule in a particularly queer way. The characteristic thing about all philosophical problems is that they arise in a peculiar way. As a way out, I can only give you examples, which if you think about them you will find the cramp relaxes.

In ordinary life one is never troubled by a gap between the sign and its application. To relieve the mental cramp it is not enough to get rid of it; you must also see why you had it. The reason which I gave for the cramp was this: that two statements that are closely connected but having different meanings are confused, a statement which is a rule and an experiential proposition. For example, to say this book agrees in color with a given sample means that in fact this book has, say, the color blue. To say the colors agree is a rule I lay down. These two statements are usually expressed in almost identical words. The point is illustrated also in the following. Suppose it is said that *A* loves *B*, meaning that he has certain feelings for *B*, but that when *B*'s life is endangered and *A* could have saved him, he did not. We say "This cannot have been love". Has the statement "*A* loves *B*" been contradicted by *A*'s not saving his life? No. It is not a contradiction to say *A* had the feeling for *B* but did not save him. It is only a conjecture that whenever *A* has a certain feeling he will do so-and-so in the future. But it is quite another thing to say I am not going to *call* this love if *A* did not save *B* when he could have done so. "If *A* had loved *B* he would have saved him" is not an experiential statement at all, but a definition or explication of what I call love. If as a matter of fact a certain feeling almost always goes together with a certain behavior, we are inclined to use feeling and behavior alternatively as criteria for love. This is all right so long as we do not get into a situation in which we have to distinguish between what we mean by love: a feeling, or behavior. These are different criteria. The same verbal expression,

"This is not love, because he does not behave as if it were" can stand for a *rule,* viz., "I do not call this love because . . .", or my saying it can mean that I do not think it is love because people do not usually behave as he did. The rule and the experiential statement are confused with each other. They of course have a definite connection, since one is conditioned by the other. But the confusion between them produces this queer mental cramp. From a distance something may look to be one thing, and be seen to be two on coming closer. Behavior and feeling are very often found together, so that we are inclined to give both phenomena the name "love" although they are different criteria. The fact that there are two entirely different criteria for having an idea, what is in the mind, and the use made of a word when we understand it, has an exact parallel in this example of the different criteria of love. It is difficult for us to survey usages of our words which are blurred in our language, and we fail to see differences that exist.

A disposition is thought of as something always there from which behavior follows. It is analogous to the structure of a machine and its behavior. There are three different statements which seem to give the meaning of "*A* loves *B*": (1) a nondispositional statement about a conscious state, i.e., feelings, (2) a statement that under certain conditions *A* will behave in such-and-such a way, (3) a dispositional statement that if some process is going on in his mind it will have the consequence that he behaves in such-and-such a way. This parallels the description of an idea, which stands either for a mental state, a set of reactions, or a state of a mechanism which has as its consequences both the behavior and certain feelings. We seem to have distinguished here three meanings for "*A* loves *B*", but this is not the case. (1), to the effect that *A* loves *B* when he has certain feelings, and (2), that he loves him when he behaves in such-and-such a way, both give meanings of the word "love". But the dispositional statement (3), referring to a mechanism, is not genuine. It gives no new meaning. Dispositional statements are always at bottom statements about a mechanism, and have the grammar of statements about a mechanism. Language uses the analogy of a machine, which constantly misleads us. In an enormous number of cases our words have the form of dispositional statements referring to a mechanism whether there is a mechanism or not. In the example about love, nobody has the slightest idea what sort of mechanism is being referred to. The dispositional statement does not tell us anything about the nature of love; it is only a way we describe it. Of the three meanings the dispositional one is the only one that is not genuine. It is actually a statement about the grammar of the word "love".

Consider understanding. If someone says he understands my order "Fetch me a plant", we shall say "understanding" may mean (1) something that happens when he says he understands, (2) the whole of what he does in response to the order. But the statement "he understands" is of the dispositional form. Although it does not refer to machinery as it seems to, what is behind the grammar of that statement is the picture of a mechanism set to react in certain ways. We think that if only we saw the machinery we should know what understanding is.

When we try to get clear about understanding (or about wishing, hoping, etc.), we ask ourselves what happens when we understand. But we are dissatisfied with descriptions of what happens. Everything we bring up, such as an image, seems irrelevant. The same is true for wishing to eat an apple, or having an idea. Images are not part of understanding, but symptoms of understanding. Nothing we could describe of our states of mind seems to be what we mean by understanding a word or sentence. It is because of the form of words, "I understand this", "I have an idea", that we suppose the grammar of these words is that of describing a state, whereas it is not. "I understand" is used quite differently. Nor does it mean that I am going to behave in a certain way, for then we have only a hypothesis. Insofar as I do not conjecture that I understand, i.e., insofar as I know that I do, understanding is an *experience,* an occurrent state. This state does not guarantee any future behavior connected with it. The question, What *is* understanding?, or What is knowing how to use a word?, is misleading. What one *can* describe is the use of the words "understanding" and "knowing".

The expressions "being able to", "understanding how to", "knowing how to go on", (for example, in a series) have practically the same grammars. When a person knows how to go on, given the series 1, 3, 7, 15, the mental states, images, etc. that occur when he knows this would not all be the same but would have resemblances or family likenesses. What happens in knowing how to go on is a vast number of things, all constituting a family. Although going on does have something to do with mental occurrences, e.g., imagining the next number in the series, 31, more than these are required as criteria for his knowing how to go on. For if he stopped after supplying one number we should probably not say he could go on. We must have certain empirical evidence. We think ourselves justified in saying he could go on if he passes certain tests, viz., if he goes on for a number of places. The fact that we are justified in saying *"A understands if A does so-and-so"* shows that the italicized sentence is a grammatical rule, just as a definition is. The same applies to "He has an idea of a plant". Knowing how to use the word "plant" is justification for saying he has the idea.

Lecture IX

What is meant by being able to continue a series? Does the statement "He can go on" mean either "He writes down the formula" or "He writes down some further digits" or both of these? Or does it mean something more? The question asks for a criterion for one set of words to mean another. Whether *he* means by "I can go on" that he sees a formula might be found out by asking him. What is meant by the word "can" here? Perhaps he means only to distinguish between seeing the formula and not seeing it. Analogously, what does the doctor who examines a man's bones and muscles mean by "He can walk"? Does this only mean that his bones and muscles are in such-and-such a state? One could say that the word "can" is used to distinguish between one state of his bones and another, if the doctor merely examines his bones to determine their condition. As far as his bones go, the doctor would say, he can walk. Similarly, as far as knowing the formula goes, he can go on with the series. It might be objected that this cannot be what is meant. For you can imagine a person who immediately on seeing the beginning of a series wrote down the formula, but when asked to go on never does go on. Knowing how to go on is never only seeing the formula. The suggestion is that "He can go on" instead means the logical product of (a) seeing the formula, and (b) that having seen it, past experience shows that he will continue the series if asked or if he tries. Analogously for walking or lifting. The doctor who says a person can lift ten pounds does not merely mean that his muscles are in order, but also that experience has shown that when his muscles are all right he will lift ten pounds if he tries.

The matter can be cleared up by imagining a language game exhibiting the use of the word "can". Let us imagine a set of primitive conditions which we can describe and survey easily. Imagine that in a tribe certain songs and poems are learned by heart and a person is said to be able to perform them if before he does so he can recapitulate them to himself inwardly. Before reciting publicly, he rehearses to himself. The use of language in the tribe will be such that the answer to "Can he recite the poem?" is always that if he succeeds in rehearsing the poem, he can; if not, not. Or suppose that as a condition of reciting it, he writes it down. *In general,* when a person can write down a poem, as a rehearsal, he succeeds in reciting it. The use of "can" in thus based on this particular fact. Does it mean then that "He can recite it" has the same meaning as "He has succeeded in writing it, and experience has shown that this usually is followed by saying it"? That is, does the meaning include the conjecture? Not if one means, as far as this language practice goes. The language practice is based on the fact that writing a poem is usually connected with reciting, and would not have been established without this fact. But this fact does not enter into the meaning of "can", unless one means by "meaning" the description of the whole practice of using the word. This is something which cannot be given, since any list will fail to be large enough to give all the uses of the word. Rehearsing a poem does not give the full meaning in the sense of the whole practice, which includes all the circumstances in which the word lives. It is this fact that makes us ask whether this is all there is to it. In the tribe described, writing the poem down distinguishes between being able to recite it and not being able to recite it. And if writing it is only intended to distinguish between these, then I should say this is what is meant in the tribe by "He can recite it". It is the rule for the use of the word "can".

The word "can" is used in an infinite variety of cases. A case can be imagined where a person is prepared to substitute for "He can walk", "I have examined everything that experience has shown to be connected with walking and found them present. As far as research goes, the leg is in a condition for walking". Also, one could say that "being able to go on in a series" means knowing the formula, so that being able to go on would distinguish between knowing and not knowing the formula. But remember that this is only part of the usage of "being able". Also remember that although being able to go on depends upon the fact that you have had a certain training which makes it very likely that you will be able to go on when you have the formula, the fact that experience shows that knowing the formula is usually followed by going on in the series does not enter into the meaning of "He can go on".

There are many cases in which I can say that a word is used in several senses and can give a definition. In giving a definition I am only giving you a use of the word in terms of other words whose uses you can take for granted, e.g., the definition of "grandfather". If I give you a definition of "being able to", would it wholly describe the use I make of the word "can"? *The uses of this word form an enormously large family.* I could describe to you a number of games with the word "can" which would all come near the actual use of the word. Suppose I draw a curve, and ten osculating circles which come near to describing the curve. This is the way in which I describe the use of the word "can". I shall give you a number of usages regulated by rules which will osculate the actual use. *There is no exact description.*

The word "can" in a great number of cases is used to refer to a *state.* Suppose we had a glass box with a ball in it, and that instead of saying "The ball is in the box" we said "The ball can be taken out of the box". This alludes to an activity without saying it is performed. Compare the ball in the box with a chemical which when heated gives off drops. When the chemical stops giving off drops we say that there are more drops to come and that these can be drawn from it. This differs from the case of the ball in the glass box where we had something we could call a state but which we could see. In the case of the chemical there is not a state which we can see when the chemical is not giving off drops. We thus have here a use of the word "can" in two different ways to describe a state, in the one case being more obviously a state than in the other.

To take another and somewhat different case, suppose a doctor on examining my muscles and finding them red says they are in perfect health. He expresses this by saying they can contract. In this instance one might say he is describing a state, but one that is hypothetical. It is not one that can be seen, as can the color of my muscles or the ball in the glass box. To say that the ability to contract is a state, only it cannot be seen, misuses language.

In which cases does "can" describe a state of affairs? We have uses of the word ranging from description of a state on up to cases where it makes no reference to a state. "He can do so-and-so" in a vast number of cases is used to describe a state of affairs. Sometimes it is used like a picture. For example, we want to say that being able to recite a poem is a state of our memory. Memory is a characteristic picture for the word "can". And when we are able immediately to continue a series, say 2, 4, 6, 8, without even seeing the formula, we also want to say that there must be something and it must be a state of the brain.

It is the same with "general idea". It is sometimes used for something before our minds, and then used in a series of cases where there is no such thing. What I have said here applies also to words like "good" and "beautiful". There is nothing identifiably in common to the states of affairs for which we use a word. There is only a number of overlapping resemblances. Our concepts are enormous families with various resemblances. One of our main philosophical troubles, which constantly recurs, is that we have such a family. We want to get clear about the use of a word, and so we hunt for something common to the instances the word applies to, even when there is hardly anything in common. What does it mean to see what is common? One can see what is common to ⬡ and ✳. But what is there in common to many reds? Here it cannot be seen; there is actually nothing. In the *Theaetetus* Socrates fails to produce a definition of "knowledge" because there is no definition giving what is common to all instances of knowledge. Because the word "knowledge" is used in all sorts of ways, any definition given will fail to apply to some cases. Similarly with the definition of number. The method of giving a definition of a word and then proceeding to other instances of its application which have very little in common is a mistaken method. We *can* show links which some cases have with others, but that is all. Furthermore, giving examples of usages is not a second-best method of giving the meaning of a word.

In naming the cardinal numbers 1, 2, 3, 4, 5, and saying "Now go on", I have given one most special case of exemplification of the concept *cardinal number*. But if you ask, "What is number?", I can exemplify that concept by cardinals, rational numbers, irrational numbers. This is exemplification of a kind. Would you say one knows what to call a number in every case? If I had given someone cardinals and rationals as examples, would he call complex numbers "numbers"? He might not. What have these in common that all are called numbers? Obviously no one thing. Here we see the different roles exemplification can play. To these there correspond as many different roles for "all", "any", and "some".

Lecture X

The question has been raised how far my method is the same as what is called description of meaning by exemplification. That sounds as if I had invented a method, a means of giving a meaning which is *just as good as* definition. The point of examining the way a word is used is not at all to provide another method of giving its meaning. When we ask on what occasions people use a word, what they say about it, what they are right to substitute for it, and in reply try to describe its use, we do so only insofar as it seems helpful in getting rid of certain philosophical troubles. We seem to be asking about the natural history of human beings. Yet you know that in some obvious sense we are not interested in natural history. Nevertheless, when I say that a word is as a matter of fact defined in such-and-such a way, or ask whether people might accept a certain definition, I seem to be talking about natural history. But it is not natural history to *invent* languages of our own, as I have done, and lay down rules for such languages as, for example, the chemists of the 19th century did with the language of chemistry. We are interested in language only insofar as it gives us trouble. I only describe the actual use of a word if this is necessary to remove some trouble we want to get rid of. Sometimes I describe its use if you have forgotten it. Sometimes I have to lay down new rules because new rules are less liable to produce confusion or because we have perhaps not thought of looking at the language we have in this light. Thus we may make use of the facts of natural history and describe the actual use of a word; or I may make up a new game for the word which departs from its actual use, in order to remind you of its use in our own language. The whole point is that I cannot tell you anything about the natural history of language, nor would it make any difference if I could. On all questions we discuss I have no opinion; and if I had, and it disagreed with one of your opinions, I would at once give it up for the sake of argument because it would be of no importance for our discussion. We constantly move in a realm where we all have the same opinions. All I can give you is a method; I cannot teach you any new truths. It is the essence of philosophy not to depend on experience, and this is what is meant by saying that philosophy is *a priori*.

One could teach philosophy solely by asking questions.

When we give a description of the use of a word we do so only so far as it seems helpful in removing certain troubles. For example, people are troubled by the assumption that there must be something common to all uses of the word "good". They say, "There is one word; therefore there must be one thing common to all its uses". Every philosophical problem typically contains one particular word or its equivalent, the word "must" or "cannot". The word "must" in the present case means that one is misled into supposing that because there is one word there must be one thing in common. One can be obsessed by a certain language form. One can think for years about a certain problem and make no progress because one never thinks of making up a new language. A philosophical trouble is an obsession, which once removed it seems impossible that it should ever have had power over us. It seems trivial.

The obsessions of philosophers vary in different ages because terminologies vary. When a terminology goes some worries may pass, only to arise again in a similar terminology. Sometimes a scientific language produces an obsession and a new language rids us of it. When dynamics first flourished it gave rise to certain obsessions which now seem obsolete. Something may play a predominant role in our language and be suddenly removed by science, e.g., the word "earth" lost its importance in the new Copernican notation. Where the old notation had given the earth a unique position, the new notation put lots of planets on the same level.* Any obsession arising from the unique position of something in our language ceases as soon as another language appears which puts that thing on a level with other things. When there was only one dynamics philosophers asked how they could reduce everything to one mechanism, and became obsessed. With the discovery of several other dynamics, the obsession disappeared.

In the case of the confusion between "is" and "equals", philosophers noted the use of "is" in "2 + 2 is (equals) 4" and "The rose is red", and went on to ask whether the rose equals, or is identical with, red, and so on. When logicians like Frege and Russell introduced the symbol "epsilon", a difference in the uses of "is" was brought out in a way which was not brought out between "is" and "equals" in our ordinary language. Our ordinary language is in tremendous flux, so that it is difficult to make distinctions in it. Their notation removed the temptation to treat different things as identical. I invented a notation to

*It might be said that Copernicus discovered certain facts about the planets, and that it was the discovery of these facts which removed the obsession about the earth and not the change from Ptolemy's notation. But the new facts might still have been expressed, in a complicated way, in Ptolemy's notation and the obsession not removed. On the other hand, the obsession might have been removed had Copernicus made up a notation with the sun as center, even though it had no application. Of course Copernicus did not *think* about notations but about planets.

get rid of the identity sign as used in "A = A", because nobody ever says a chair is a chair; and the difficulties connected with this use vanish.

Treating language as we have here brings with it the puzzle: What becomes of the rigidity of logic? We have the impression that logic is not a thing within our control. There seems to be a way of explaining why it is not by thinking that logicians make up an ideal language to which our normal languages only approximate. I once said that logic describes the use of language in a vacuum. Games or languages which we make up with stated rules one might call ideal languages, but this is a bad description since they are not ideal in the sense of being "better". They serve one purpose, to make comparisons. They can be put beside actual languages so as to enable us to see certain features in them and by this means to get rid of certain difficulties. Suppose I make up a language in which "is" has two meanings. Is it better? Not from the practical point of view. No ordinary person mixes up the meaning of "is" in "The rose is red" and "2 + 2 = 4". It is ideal in the sense of having simply statable rules. Its only point is to get rid of certain obsessions—it does not do more. One might of course suggest a notation or language which would be better for some practical purpose, but this would be accidental. It is not part of our design.

It is characteristic of obsessions that they are not recognized and at certain stages are not even recognizable. These are attacked as scientific problems are, and are treated perfectly hoplessly, as if we had to find out something new. The problems do not appear to concern questions about language but rather questions of fact of which we do not yet know enough. It is for this reason that you are constantly tempted to think I am giving you some information, and that you expect from me a theory. In using the words "I think so-and-so" it looks as if I were discussing the problems of a science called metaphysics.

I shall now go on to discuss propositions. All sorts of definitions of "proposition" have been given. But when I am asked about propositions I explain by examples. The examples are usually sentences in some known language, and that produces the strong feeling of there being something common to all such sentences. Some people have said that a substantive and verb are what is common. Hardly anyone will give as an example of a proposition a command such as "Take this chair to Mr. Smith" or "Walk in this direction", uttered in conjunction with a gesture. (Is the gesture combined with the command part of the command?) The proposition is usually considered to consist of words. But where there are no words, say in a line drawn like this, ⎍⎍, a person might understand that he is to walk in a certain direction. Is this a sentence? If it is, then we would not say that what is common to sentences is a substantive and verb.

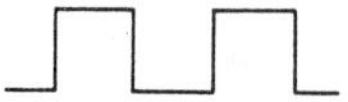

Logicians have had the obsession that the life of a proposition is the copula "is". But they know as well as we do that not all English sentences have a subject, a copula, and an adjective. They have said that every sentence can be "reduced" to such a sentence. The fact that it can be reduced is analogous to the following: that every closed curve is said to be a circle. To the objection that a certain curve is not a circle, the reply is that it could be projected into a circle. To twist "Every closed curve can be projected into a cricle" into "Every closed curve is a circle" is exactly paralleled by "Every proposition can be transformed into one of the subject-predicate form". But in neither case have I said anything about the method of projection, and until I do, I have said nothing. I could have said that I was going to use a symbolism in which every closed curve would be represented by a circle. That is a *rule:* to replace "closed curve" by "circle". Similarly, I could say that I would transform every propositional form into a subject-predicate sentence. The statement that every proposition has a subject and predicate could assert a fact of natural history; there could be a language composed entirely of such sentences. On the other hand, in uttering that statement I might be laying down a rule, a rule about the jingle of my sentences. But the rule does not enlighten us unless I say how I propose to transform all propositions into this form, i.e., what is to be done with other words than substantives, copula, and adjectives.

I want to say something about the family which we call propositions. I want to give you some idea of its multitude, and to make the point that there is little chance of giving a definition which will cover them all. {The family of propositions has many things in common and many things not in common. What most have in common is assumed to be common to all. A vast number of propositions have in common the propositional *form* of a given language—the jingle. And this feature is taken to be a common feature of all. Sentences read backwards would as a matter of fact not have the jingle. But of course there may be propositions which do not have this jingle at all.} *

* Portion in braces taken from The Yellow Book. (Editor)

In logic we talk of a proposition as that which is true or false, or as that which can be negated. And we have a calculus of propositions. I want to discuss one notion occurring in this calculus, negation, and to show first the family of negations. "Negation" has different uses. In the logical calculus $\sim\sim p = p$. What sort of proposition is this? One is tempted to say negation behaves in accordance with this law, and there is some truth and some falsity in this. The question has been raised whether it is a rule or whether it is a statement about our habits of using the negation sign. It is not a statement about our habits, for then it would be a statement of natural history and is not even true. To an uneducated person a double negation means a denial, not an affirmation, e.g., "He don't know nothing about it". In order to get clear about propositions we shall need to go into more detail than is usually done. This will require showing the family of propositions and what I call the family of negations. You will constantly be asking whether "not" has different meanigs. "$\sim p$" may symbolize not-*p* in two symbolisms, and yet in the one "$\sim\sim p$" may symbolize *p* and in the other not-*p*. One can fall into confusion by asking whether "$\sim\sim p$" and "*p*" mean the same.

Lecture XI

Difficulty is created by the fact that we have invented an enormously complicated language for our use and we are all grown-ups. The philosophy of a child would be quite different from ours, but in a different sense than the physics of a child would be. The physics of a child would be different because it does not know various physical facts, but its philosophy would be different because its language is simpler. It will be very valuable to study more primitive examples of language, what I call "language games" (synonymous with "primitive languages" for the most part). These will bear the same relation to our language as primitive arithmetic bears to our arithmetic. It is a fallacy to suppose these languages are incomplete. Primitive arithmetic is not incomplete, even one in which there are only the first five numerals; and our arithmetic is not more complete. Would chess be incomplete if we knew another game which somehow incorporated chess? It would be merely a different game. To think otherwise is to confuse mathematics with a natural science. If mathematics were a science of numbers as pomology is a science of apples, then a mathematics which did not include irrational numbers or numbers after 5 would be counted incomplete, just as would a treatise on pomology which left out reference to one sort of apples. And the latter would be incorrect if it invented kinds of apples which did not exist. But mathematics is not a natural science.

Surrounding the blurred whole of ordinary language there are the special languages, e.g., the languages of chemistry and meteorology. I shall consider a language as such a conglomerate. In our language we find a mixture of descriptions, hypotheses, questions, orders, etc., but any list we made of these would be entirely inadequate. Let us compare it with a simple language in a tribe in which only orders are given. We, who talk about the tribe having this language, call these "orders" because the rule these words play in the life of the tribe is that of orders. The word "order" is not in their language, nor is there any such thing as conversation. The whole object is communication between a builder and his workman. The builder orders "Brick!", for example, upon which the workman brings him a brick.

We shall suppose that a child learns this language by being drilled. He is given, say, ten words, such as "brick", "column", "clay". In the description of this training is understanding left out? You will say the child must understand the words else he cannot be taught to react to orders. I reply, Certainly, if you like, just as a dog can be taught to look after sheep. A calf or cat cannot be taught; I could go through all the motions with these animals and would not get an appropriate reaction. Training can be described as consisting of two steps (1) the trainer's doing certain things, (2) the occurrence of certain reactions on the part of the subject, with the possibility of improvement. Teaching a language always depends on a training which presupposes that the subject reacts. If the subject does not react in a given case, that is, does not understand, reference to understanding will then not appear in the description of the training. But nothing is omitted from the description by omitting reference to understanding.

Now there is a certain preliminary exercise to obeying the order, namely, learning what to do when an order, e.g., "Brick!", is given. This is very close to what we should call "giving the thing a name". The mother puts a brick on a pile and says "brick", and then the child does the same thing. Notice that "brick", said in the presence of the child is not properly an ostensive definition, because in this language we have not yet the question, What is this called? It is a process of naming in a different kind of surroundings.

The question might be raised whether the word "brick" has the same meaning in this language as in ours. You might say that the builder means by it what we mean by "Bring me a brick". But this would be dangerous. Although these expressions play the same role in the two languages, in the primitive language the words "bring me" do not come in. We could imagine that even in English, although we said "Bring me so-and-so" for everything else, instead of "Bring me a brick" we said "Brick", as in the orders "Charge" and "Fire" in military usage. Then the word "brick" would play a different role from what it plays in the sentence "There is a brick".

Consider now another language (2) in which an order consists of two words. Besides the words "brick", "column", etc. we have a series of letters A—J or a series of ten notes, say the first ten notes of "God Save the King". These must be learned by heart, whereas words such as "brick" are not. The order now consists of a word and a letter, say "E brick!". The child must go to the pile of bricks, take up one brick for each letter through E, and bring E number of bricks to the master. The letters of the alphabet are thus seen to be numerals in this language. The tribe has a very primitive arithmetic in which it can count up to 10, but has no addition or multiplication. Note how different are the functions of the words of this language (a) that of the letters of the list of ten, which must be learned by heart, and (b) that connected with actions of bringing something which the builder orders. Although the word "E" and "brick", as spoken or written, are similar, their functions are in no way comparable.

You will notice that in the two languages which we have described there is in a particular sense no "understanding". There is nothing corresponding to asking for the name of a thing or giving it a name. The philosophical question about meaning would not arise for the philosophers of our tribe.

We now introduce another game (3) having question and answer. We might have, say, twenty-five letters or numerals, and the words "brick", etc., as before. We suppose the helper can count the bricks against his letters, and that for any number beyond twenty-five he says "many". The role of "many" then is rather like that of a numeral and yet different. The question in our game might always be "How many?" This would be answered by "J", or by "Y", or by "many" for a number over twenty-five.

In the game (1) we had in the training something which was somewhat similar to ostensive definition. For the numerals in game (2) we could have a *sort* of ostensive definition. When shown three bricks the helper would be taught to say "C" or "3", instead of learning the numbers by heart, and this might be an ostensive definition of "3". Here we have a different sense of ostensive definition. Three columns would be "3" as well as three bricks.

Another language (4) might introduce the word "there", which has a different function still from substantives and numerals. An order would be, for example, "J bricks there!" together with a gesture of pointing, which would be followed by the helper's putting bricks over there. Look now at the use of the word "there". One might perhaps say it is the name of a place. But to call it a name is to use the word "name" in a very different sense from that of the name "Charing X". "There" has no meaning unless it is accompanied by a gesture. Are expressions such as "Brick!" and "J bricks there!" sentences? As you like. You can draw the distinction wherever you like, but it is not easy to show why it should be drawn at any particular point.

Many other games can be made up, for example, one which introduces playing cards for which we invent names and uses. Or one which introduces the question as to what a thing is called, and the answer to that question. Again, we could make up a game involving the use of a table, on one side of which are pictures of a house, table, ball, etc., and on the other side some words. The child is trained to go from a word to a picture, and then to bring the article of which it is a picture. We might then leave out one sign and have the child supply it and use it as he uses the other signs. Again, we might have a game which introduced descriptions and proper names. The latter have a unique function. If a child is trained to call one brick "Jack", we must have a means of identifying this brick and of following its movements to a different place. Suppose you hold two bricks in front of you, and pointing to one say, "This is Jack". Now change them behind your back and present them again. How is the child to know which is Jack unless he has followed your movements? Questions about proper names are enormously more complicated than some logicians suppose.

To make up a game having descriptions we might introduce a different kind of sign, consisting of words like "5", "brick", and "above", "below", "right", "left". A description will be, say "5 bricks left", or "6 bricks below". The description will have a particular jingle, and we can make other combinations having the same jingle. Some combinations will make sense and some will not. We can now introduce the words "true" and "false". A game could be made up in which a description is given and someone builds something in accordance with it, whereupon the supervisor says "true", or "false". Or we might have a game in which we say "true" when someone counts to a certain number and "false" if he misses out a number. You may object that these are not examples of being true, or false; but I say that is one way in which the words can be used.

Do not make the mistake of supposing that I am showing how language is built up or how it has evolved. Sometimes it is easier to imagine these invented languages as languages of a primitive tribe and sometimes as the actual primitive language of a child. A child does actually begin with such a primitive language. Its language training is mostly in the form of such games. A new game introduces a new element into language, for example, negation. It will be noticed that the elements we have already introduced are of great variety. The difficulty of this method of exhibiting language games is that you think it is perfectly trivial. You do not see its importance.

Lecture XII

People have been extremely worried about the idea of negation, and have tried to say that "not. . . ." is really a disjunction. Is the proposition not-*p* the same as *r or s or t or . . .*? Sometimes it is, and sometimes it is not. As an example of the first alternative, consider the order: "Bring me a primary color but not yellow." This comes to: "Bring me red, or green, or blue, or black, or white." Negation and the corresponding disjunction are the same here, as "primary color" is defined by an enumeration. This is not the case with the order, "Paint me a color but not this red", or with the statement "He is in the house but not here". The order, "Write the permutations of *a, b,* and *c* but not *acb*", is like the example about primary colors. In most cases we can tell whether a negation is a disjunction or not. If we can say what the disjunction is, then negation is disjunction, if not, not. Whether or not there is a definite number of alternatives that could be given, we do not always regard negation as disjunction. For instance, if there were 123 species of mammals, this does not mean that for "mammals but not whales" a disjunction could be substituted. Note that the statement that there are 123 species of mammals is experiential, whereas "There are six permutations of *a, b, c*" is not. The latter is a grammatical proposition, a rule about the use of the word "permutation". "Is a permutation of *a, b, c*" *means* "is either *abc* or *acb* or *bac* or *bca* or *cab* or *cba*''. On the other hand, were it true that lions, rabbits, and dogs were the only species of animals in existence, ''I saw a lion, rabbit, and dog'' is a different proposition from ''I saw one of each the species of animals on the earth''. By contrast, ''I saw a lion, rabbit, and dog in all possible permutations'' is the same as ''I saw *this* permutation of a lion, rabbit, and dog, and *this* permutation. . . . and this one'' (through six). And ''I saw four animals'' is the same as ''I saw the sum of $2+2$ animals''. The language of mathematics which comes into these statements functions as a bit of grammar which adds nothing to them.

The following are quite different cases where negation is not a disjunction: ''Write down the roots of the equation $x^2+3x=4$, but not the negative one''. This is equivalent to the positive assertion, ''Write down the positive root''. And ''Write down a cardinal number, but not 3'' is for a different reason not a disjunction. Here you could not state a disjunction of all the other cardinals, and it is nonsense to give as the reason that you do not have time. That there is an infinite number of cardinals is a rule you make, not an empirical proposition. You will have to say that in this game there is no end, so that to write down or not to write down the cardinals other than 3 will equally be nonsense.

The words ''true'' and ''false'' are two words on which philosophy has turned, and it is very important to see that philosophy always turns upon nonsensical questions. Discussion of these words is made easier once it is realized that the words ''true'' and ''false'' can be done away with altogether. Instead of saying ''*p* is true'' we shall say ''*p*'', and instead of ''*p* is false'', ''not-*p*''. That is, instead of the notions of *truth* and *falsity*, we use *proposition* and *negation*. That we can do this is a useful hint, but it does not do away with the puzzles connected with truth and falsity.*

Let us examine the statement that a proposition is true if it agrees with reality and false if it does not. We must look at language games to see what this agreement and disagreement consist in. There are cases where what is meant by agreement and disagreement is clear. Let us consider a game in which descriptions of things are given in the form of drawings. Where drawing and original are alike we know

*On the other hand, we could do away with *negation*, *disjunction*, *conjunction*, etc. and use *true* and *false*, making up a notation containing only the words ''true'' and ''false''. I once did that, with the notation for truth-functions. By replacing our ordinary notation by this one, what logical propositions are is made clear.

what is meant by agreeing. Likeness is a most common form of agreement. But a drawing which is out of perspective might also be said to agree. That is, a thing might not be like reality and be said to agree with reality, for example, a painted picture, a statue, a picture drawn according to queer rules of projection, a map. So if we say a proposition is true when it agrees with reality we must say in what way it agrees with reality, since the expression "agrees with" is used in all sorts of different ways. To say that *p* is true if it agrees with reality does not say as much as it seems to say, though it might be useful to say this provided we have an idea of agreement which we have not got of truth.

Consider the idea of projection. There are the same difficulties with this idea as with the ideas of agreement and of truth. Imagine a language consisting of four letters whose meaning is given in a table coordinating them with arrows.

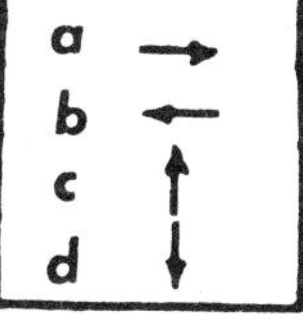

A letter or combination of letters gives an order to move according to the arrows, for example, *aabdc*. A representation by arrows

could be called a projection of the letters, and could even be called a picture. Starting with a likeness, by extension one can get to something very unlike. For example, all pieces of furniture could be looked on, by extension, as chairs. Suppose I said that everything in this room is a chair, and that someone objected that there are tables, a door, etc. I could reply that on a scale of likeness a stool is between a table and a chair, and a stool is a kind of chair, this small table is like a stool, this large table like this small table, a door like the top of the table, and so on. There is an enormous temptation to regard all things as extensions of something else. This is the sort of temptation we fall into in saying a proposition is true when it agrees with reality. We stretch the idea of agreement as I have stretched the likeness of a chair to other pieces of furniture.

Imagine a room having nothing in it but what are usually called chairs. To say there were only chairs in it would be a straightforward statement of fact. Now compare my saying the same thing of the present room, and answering objections by: "But this (an ottoman) is only a chair without legs, this (a table) is a chair with a nonexistent back", etc. You could reply that to say there are nothing but chairs here is to say nothing; for whatever is in this room, I could have come to that statement by extending or compressing the notion of a chair. I have drawn no limit, and that statement gives no clue as to what is a chair or what is in the room. Now I could do something different, which *might be useful:* adopt a notation in which everything in the room is to be described by giving its deviation from a chair. Note that the statement, "Each thing in the room deviates in some way or other from a chair", is not a statement about the nature of things in this room, but a grammatical statement about a description I wish to adopt*. Here we have a rule. Instead of saying "This is a table", we would say "This piece of furniture deviates from a chair in such-and-such a way". It might be important to stress the similarity. To say that there is agreement between a proposition and reality is to say nothing because we do not know what is meant by agreement. But we could give language games [such as the description of things by their deviation from a chair] showing the idea of agreement or extensions of it.

There is one peculiar difficulty about the ideas of negation, truth, falsity, proposition, which is expressed in this crude form: that a proposition is false or its negation true when no fact corresponds to the proposition. But if no fact corresponds to it, why is it not nonsensical, as a name would be if it did not name anything?

As in the case of every philosophical problem, this puzzle arises from an obsession. Philosophy may start from common sense but it cannot remain common sense. As a matter of fact philosophy cannot start from common sense because the business of philosophy is to rid one of those puzzles which do not arise for common sense. No philosopher lacks common sense in ordinary life. So philosophers should not attempt to present the idealistic or solipsistic positions, for example, as though they were absurd—by pointing out to a person who puts forward these positions that he does not really wonder whether the beef is real or whether it is an idea in his mind, whether his wife is real or whether only he is real. Of course he does not, and it is not a proper objection. You must not try to avoid a philosophical problem by ap-

*I once said that a proposition is a picture of reality. This might introduce a very useful way of looking at it, but it is nothing else than saying, I want to look at it as a picture.

pealing to common sense; instead, present it as it arises with most power. You must allow yourself to be dragged into the mire, and get out of it. Philosophy can be said to consist of three activities: to see the commonsense answer, to get yourself so deeply into the problem that the commonsense answer is unbearable, and to get from that situation back to the commonsense answer. But the commonsense answer in itself is no solution; everyone knows it. One must not in philosophy attempt to short-circuit problems.

Let us allow ourselves to be dragged into the mire in connection with problems about false propositions. Take the sentence "There is a human-headed chair in this room". This makes sense but is untrue. We might say that in a sense nothing corresponds to it. What is its connection with reality? What prevents it from being nonsense? One *might* answer, "It is not nonsense because to each of the constituents of the proposition something in the room corresponds, although nothing corresponds to their combination". That is, things are not arranged in the way the proposition says they are arranged. At first sight this seems a good answer, though on further consideration it does not. One trouble with it appears in the questions: "What are the constituents of the proposition?", "What are the constituents of the fact which correspond, or fail to correspond, with those of the proposition?" It is all very well to say that the constituents of the proposition correspond with the constituents of the fact, but what are the constituents? Are they legs, back, seat, head, etc., or are they atoms, or are they color, shape, etc.? Some people say we do not know what the constituents are, but that this is a matter for further analysis. Compare Russell's individuals. Another trouble is that it does not help to say that the constituents are not combined in reality as they are asserted to be combined. You have only said that the sentence contains several words which have meaning but that the whole combination of words does not correspond to anything. So the problem remains.

The difficulty is that we want this proposition to be false and true at the same time. "There is a human-headed chair" has nothing corresponding to it, and yet we think there must be something corresponding to it, a sort of shadow of reality. But we are no better off; the shadow gives the same trouble all over again. For why on earth should it be a shadow of *this* reality? The puzzle about negation is in the idea that something must correspond to a symbol.

If we are in the mire, a specially chosen example may immediately pull us out. On looking at the way a symbol is used, e.g., the way in which our arrow signs, or letters, are used to describe a route, the difficulty we are in does not seem to be present at all. It occurs at once if the idea of meaning, or of knowing what is meant, is brought in. I could describe the use of the symbol *aabcc* by describing the way a person goes; but what does it mean in the case where he does not go in the way described? Puzzles arise when one tries to fix a meaning for such phrases as "knowing what an expression means", by reference to something that corresponds to the expression, especially when it asserts what is not the case. How can one know what is meant when nothing corresponds to it? Yet one must know what one means when one says there is a human-headed chair here. A grammatical obsession can be described as taking some extremely simple form of grammar and so to speak conjugating all words according to its pattern.

If we look at the way we use the word "correspond" and the way we use "knowing what so-and-so means", that will probably be sufficient to clear up the difficulty. To know what the sign *aac* means may consist in drawing a figure

or in using it correctly, say, by walking the right path. We might draw a line for the path and call the line a shadow of the way one goes. It might be said that in order to carry out the order "*aac*" one must understand it, and understanding the order might consist in drawing a plan. But then must one understand what the plan means? There seems to be an infinite regress, that there is no stopping anywhere. It seems that if we stop short we do not understand the order. But this is what we do call understanding.

Lecture XIII

Thinking, wishing, hoping, believing, and negation all have something in common. The same sort of puzzling questions can be asked about each: How can one wish for a thing that does not happen or hope that something will happen that does not? How can not-*p* negate *p*, when *p* may not be the case, i.e., when nothing corresponds to *p*? To the latter question I have pointed out one possible answer: that what corresponds to the negation of *p*, although not the fact, are the constituents of the fact. For example, corresponding to "There is no chair here" there is the place here, and there are chairs in the world. A similar thing can be said of wishing for something which does not happen, e.g., that Smith would come into the room and he does not come in: that the constituents of this fact exist but are not combined as are the constituents of the wish. When I say I know that Smith shoots with bow and arrow, the fact that he does so seems to come into the fact that I know he does. Similarly, when I wish that he come into the room and in fact he does not, we think that that fact or some shadow of it must somehow enter into my wishing. The problem as to the constituents of the wish is the problem concerning part of a complex sign, taken separately. Just as a box consists of a bottom and lid, we suppose that a proposition such as "I wish Smith would come" must consist of constituents put together in some way. We wrongly compare this proposition with a box consisting of parts, "I wish that" and "Smith comes" being the different parts.

Suppose I said that wishing for the box presupposes imagining a box. I do not wish for the image of the box, for I have that, as the constituent of the wish; but do I not wish for something similar to it? This tangle can be undone by destroying the idea of similarity. For could not wishing for this box be imagining some strange projection of the box? What we seem to want is that the thing wished for be the same as, not something similar to, the fulfillment of the wish. We do not want a copy. A gap between wish and fulfillment will not do.

If "$\sim p$" is understood, then "p" must also be understood. But if p if false, then nothing corresponds to it. We know what it means even though it is not true; but what is it to understand it, or know what it means? For example, what does it mean to understand the order "Leave the room", when you do not leave the room? Your understanding may be a picture of your leaving the room, but of course that is not the same as leaving the room. The order given in words might be translated into a picture, but it won't do to say that understanding the order consists of nothing more than translating the words into a visual image or picture. If that were all, one might say you did not understand the order; you were not ordered to have an image or make a picture but to leave the room. You have not carried out the order, and are no nearer to carrying out the order by making a picture. It is as if understanding ought to have taken you up to the point of carrying out the order. Yet one does not mean that you should have carried it out, for it may be understood without being carried out. The difficulty would disappear if a class of cases of understanding orders was correlated with a class of acts of carrying them out.

Suppose we order someone to do certain physical exercises, and then immediately show a film of such exercises. Now understanding the order, if it means seeing these moving pictures, does not include carrying out the order. Understanding the order, although necessary for carrying it out, cannot anticipate the execution of the order. And not being able to anticipate the execution, it seems that understanding cannot do what it ought to do. We are confronted by two simple facts, that understanding an order and carrying it out are different, and that in the sentence, "Smith understands he is to leave the room", the whole order appears. Understanding the order does not involve execution, which is what the order refers to, so part of the sentence "Smith understands he is to leave the room" seems superfluous. Understanding the order seems to stop short, because it is not the execution. Why talk of understanding the order if it cannot comprise its execution? But there is no redundancy in sentences such as this one. For the process of carrying out the order and the understanding of the order have the same multiplicity. This is the important factor, not the fact that they are similar. We have three systems, (1) the system of verbal expressions, (2) the system of pictures, (3) the system of actions. All three have the same multiplicity. It is not necessary that the second be the same as the third in order that there be understanding, but instances of (2) and (3) must have the same multiplicity. Hence nothing is redundant. We have here projective relations leading from the words to the pictures and from the pictures to the actions.

I could give a sentence "p" certain indices, "p'" and "p''". Let "p" be the sentence "Smith leaves the room", and let "p'" stand for the picture of Smith leaving the room, and "p''" for the action performed by Smith. These all have the same multiplicity. "Smith understands" corresponds to the first index, and "Smith carries out the order" (by leaving the room) to the second index. To every act of utterance a process of understanding will correspond, and to every variation in understanding the order a variation in carrying it out will correspond. The description of the carrying out will only differ from the description of the understanding by an index.

Suppose I give a negative order, e.g., "Don't draw a line through this circle". In the order there appears the description, "drawing a line through the circle", which refers to what is not to exist. Let us translate the order into picture symbolism

In this symbolism you will see that the picture does represent understanding the order. Thus, "Do not draw a line through the circle" will be represented by a circular figure with a line drawn through it. Does this mean that in the understanding of the order there is contained the carrying out of the opposite order? Here we need the index "not", meaning "to be avoided". It is an index accompanying "Draw a line through the circle", just as "understand" was the index accompanying "Smith leaves the room".

In both pairs of sentences, (a) "Smith leaves the room" and "Smith understands he is to leave the room" and (b) "Draw a line through the circle" and "Don't draw a line through the circle", the same subordinate sentence appears as constituent. And you think the same fact must occur as constituent. But nothing of the sort may occur. The mistake which leads us into philosophical trouble is supposing that just as a subordinate sentence can appear in the sentence, so a fact corresponds to this constituent as to the whole. "Not" and "understanding" are only indices and can change the whole way in which a sentence is used. When I have a sentence "*p*" and add an index such as "not" or "understand", all it may describe is a projective relation. To understand "*p*" is not like doing something described by "*p*". Similarly for sentences beginning with "wish". The sentence "I eat an apple" and "I wish to eat an apple" are entirely different. But it is wrong to explain why sentences so different use the same words by saying that the constituents of one are included in the other, or that "I eat an apple" and "I did not eat an apple" differ in having a different arrangement of the same constituents. In "not-*Fx*" "not" is an index which changes the way in which *Fx* is used. "not-*Fx*" expresses a projective relation.

Lecture XIV

The characteristic of words like "understand" and "can" is that they are used alternately for (a) something occurring in the mind as a conscious event, (b) a disposition, and (c) a translation. The use of "understand" in case (a) is illustrated when one says "Now I understand". Its use in case (b) overlaps with "is able to", and is illustrated by one's being able to do a certain thing when one understands. The two overlap in a special way where understanding is the same as being able to use a sign. "Understanding" is used in case (c) when a translation into a picture or other symbolism is involved, as in understanding a word or sentence to mean so-and-so as contrasted with merely understanding it, or in understanding an order by visualizing a film of its execution. Problems about understanding are problems caused by mixing up these meanings. This is not to say its use is never clear.

The word "understanding" is used to distinguish understanding from not understanding in various circumstances: (1) After hearing something without paying attention one can understand it when one realizes what happened in the margin of consciousness or when half asleep. (2) Or one can understand a complicated English sentence when one is able to see how it is or ought to be punctuated, what a pronoun refers to, how dependent clauses are to be separated out. (Try understanding the Explosives Act on any railway station!) (3) Or one can understand something when one is able to visualize what one could not visualize before. (4) Or a child understands when he acts in a way he did not before. [The difference between understanding and not understanding is often clear in these cases. Not understanding, like understanding, also occurs in quite different circumstances.] Not understanding one word, like "grapefruit", is different from not understanding a French or Russian sentence. And not understanding the word "undulates" in "He undulates his hair" is different from not understanding the word "hair" as well.

Some people say that understanding a sentence consists in the impression made by every word. (Compare William James, who said that special feelings attach to "if . . . , then . . .", "but", "and"). This sounds like a simple statement but is really extremely complicated. I could of course say there are sensations attached to "if . . . , then . . .", etc. There are the sounds of the words, and all sorts of bodily sensations connected with gesture and intonation. Where we are liable to go wrong is in supposing that sensations connected with words are somehow "in the mind". The phrase "in the mind" has caused more confusion than almost any other in philosophy. Sensations need not always be present when the words are uttered. It is not wrong to say there are bodily sensations accompanying a word so long as you do not say these sensations must be there whenever you say the word or understand the sentence in which it occurs.

To see what is meant by "understanding" it is useful to consider understanding a picture. Suppose you are incapable of seeing the patches on a canvas as forming the surfaces of bodies, that they are seen as patches in a plane. This could be called not understanding, and seeing them in three dimensions might be called understanding. If these bodies are not recognized as anything you know, as contrasted with their being recognized as men sitting in chairs, you would probably still say you do not understand. Consider another case: seeing the picture of chairs and people sitting in them without its making any sense to you, and then suddenly seeing it as a classroom with pupils and a teacher. Thereupon you say, "Now I understand". The word "understanding" here means understanding an impression. It is like understanding the punctuation of a sentence, or like understanding an isolated sentence such as "After saying this, he left her". In a sense I understand this sentence but in a sense I do not. I suppose it means that a man was talking to a woman and then went away. Suppose this were the first sentence in a book. After reading the book I might say, "Now I understand the first sentence", because I know who they are, etc. Yet I had perhaps not read it again.

The trouble with words such as "understanding" comes through thinking of a few cases and trying to carry over their analogy to all other cases. For example, conscious mental acts do play a great role in understanding, but we should not try to make every case of understanding look like these cases. For there are cases where no conscious experience mediates between understanding an order, say, and carrying it out. Nor should we construe what we cannot do in mathematics after the case of human frailty. Troubles we get into in philosophy come through constantly trying to construe everything in accordance with one paradigm or model. Philosophy we might say arises out of certain prejudices. The words "must" and "cannot" are typical words exhibiting these prejudices. They are prejudices in favor of certain grammatical forms.

Let us turn to negation as it occurs in mathematical and empirical propositions. Does "negation" have the same meaning in "He did not leave the room" as in "$2+2 \neq 5$"? Some people say the two have the same meaning because it is not true that he left the room and not true that 2 plus 2 make 5. But nothing at all is explained by this translation, since "it is not true that so-and-so" is only another way of saying "not so-and-so". If someone justifies this as an explanation by saying there is the same feeling in both cases, then there is no more to be said if that is the criterion. What is similar or different in these two kinds of cases is the grammar of the word "not". If "not" is so used that "not-not-*p*" is equivalent to "*p*", or "not-*p* or *q*" to "p implies *q*", this gives the grammar. By producing such rules you can show in what respects the grammar of experiential and mathematical propositions is the same. There are, of course, some respects in which negation is the same in both. For example, it might be said that negation in experiential propositions is more like negation in mathematics than it is like conjunction. On the other hand, you will also find a large number of rules which do not apply, so you can please yourself whether you say "not" is different in the two cases.

Consider "The earth does not move in a circle around the sun" and "I do not have toothache". These are utterly different negations. In the first case there are certain observations which bear out the hypothesis about the movement of the earth, but there is no such thing as my confirming my toothache. These two propositions might both be called experiential propositions, but they are utterly different instruments. Even inside mathematics negation plays different roles, as do "all" and "any". There is a vast difference between talking about a cardinal number that is not 4 and a real number that is not 4. There are no fewer uses of the word "not" than there are of the words "all" and "any". The expression of generality covers a vast number of different uses, and it is from this fact that many of the confusions in the foundations of mathematics arise.

If I say that "not", "all", "some", "any" have different grammars when used in connection with cardinal and real numbers, this must come to saying that the word "proof", among other words, also has different meanings as applied to cardinals and reals, for example, the proof that there is a cardinal number fulfilling certain conditions, in contrast to the proof that there is a certain real number fulfilling them. One could say "proof" has as many different meanings as there are proofs. All the proofs form a family, and the word "proof" does not refer to any one characteristic of those processes called proofs. Bringing together each single proof with others would make one see that the family of proofs is not of the same kind as the family of apples. Proofs form a family in this way: some are closely connected, as are a series of multiplications, some are less closely connected, as addition and x by dx. Proofs in Euclid are another family. No feature of a proof is irrelevant. There are proofs in connection with which there is a rule for making up similar proofs, e.g., for proving that a certain number is a multiple of two others. But in Euclid there are no such rules; each proof is a sort of trick.

Weyl said that every existential proof must consist in constructing what is said to exist. But *must* it? Doesn't this depend on what is called an existential proof? Weyl is using the fact that in a huge number of cases something is done which might be called constructing a certain entity. What is an existential theorem? The answer is this, and this, and this . . . If there were such a thing as existence which is proved when an existence theorem is proved, then perhaps one could say every existential proof must do a certain thing. Weyl talks as though he has a clear idea of existence independent of proof, and has made what looks like a statement about the natural history of proofs in saying that only such-and-such prove existence. There is no concept of existential theorems except through the special existential theorems. Every existential proof is different, and "existential theorem" has different meanings according as what is said to exist is, or is not, constructed. Of course one can arbitrarily fix a criterion: one can call an existential proof one which fulfills certain formal conditions.

What is the concept of *number?* Suppose the commutative, associative, and distributive laws are taken as criteria for something's being number. This is to define number in a formal way, as whatever obeys these rules. But we do use "number" for things for which these laws do not hold, and there are transitional cases. We call cardinal numbers, irrationals, and real numbers all "numbers". But these have utterly different grammars, and to say we cannot make the same statements about cardinals as we make about real numbers is like saying we cannot use a chessboard in whist or a net in Rugby. There is enough in common between cardinals and reals, namely these laws, to make us call them both numbers, just as chess and draughts have something in common. But they are entirely different games. We call cardinals and irrationals numbers because in certain respects they are analogous, although they differ in other respects.

We could have a perfectly good arithmetic the numbers of which were 1, 2, 3, 4, 5, and *many*. This is analogous to our arithmetic with 1, 2, 3, 4, 5, . . . It is misleading to say that with ours we cannot reach the end of the numbers, whereas in the other arithmetic we can. The word "can" makes this statement look like an experiential proposition such as "He can lift 100 pounds, while no human being can lift 1000 pounds". But we have made up the grammar of our numerals in such a way that there is no end. We have provided no end. Compare the two arithmetics with a game played within a court and the same game played without any boundaries. What is it that is infinite in the latter game? Not the physical field. Rather, the *rules* give an unlimited allowance for the size of the playing field. It is silly to say one cannot reach the limit when the rules provide no limit.

Similarly, the difference in the use of "not" as applied to cardinal and real numbers is to be found by comparing the grammars of cardinals and reals. What cannot be done with cardinals that is done with reals is not to be put down to human frailty, any more than the fact that we cannot count all the cardinals.

Lent Term, 1935

Lecture I

We were discussing last term language games and primitive kinds of negation. Why do people ask whether a negation is equivalent to a disjunction? We might say that they ask the question because they wish it to be so; they want to do away with negation because they feel that there is something queer about it. The queerness is that when not-*p* is true what is negated is not the case. But negation is only queer because it it looked at in a certain light. We might ask why philosophy deals with such questions. I could give you a psychological and historical reason. There is no trouble at all with primitive languages about concrete objects. Talk about a chair and a human body and all is well; talk about negation and the human mind and things begin to look queer. A substantive in language is used primarily for a physical body, and a verb for the movement of such a body. This is the simplest application of language, and this fact is immensely important. When we have difficulty with the grammar of our language we take certain primitive schemas and try to give them wider application than is possible. We might say it is the whole of philosophy to realize that there is no more difficulty about time than there is about this chair.

The word "negation" does not look queer until we use it in connection with the word "exist" or some word having the same grammar as "exist". When we say "This chair is not green" we seem to be referring to a fact that does not exist. The method of dealing with negation which I shall use here is to give the queer aspect and then gradually change over to one that is not queer. There is no reason for doing away with negation. It is no solution of the problem to make of a negation the disjunction "this or this or . . . but not that". The idea of solving a philosophical problem in this way is absurd.

We say of the word "yellow" that if it is to have meaning there must be something yellow somewhere. But why this "must"? Could not everything yellow have been destroyed? Suppose you learned the names of colors from a chart which correlated colored patches with certain words like "yellow", "green", etc. It is not necessary that if one is to understand the word "orange" something orange must exist. And if we have a game in which the sample is orange, then it is nonsense to cite the sample in substantiation of the claim that something orange must exist. This would by like saying that there must be something a foot long because the Greenwich foot, the paradigm, is a foot long. Or like saying that in order to speak about five things there must be five things, where the latter, five letters, say, are the paradigm.

The question is, What does one *do* with a negated sentence such as "This is not green"; how is it used? We can construct many usages. In philosophy, sentences like "This is not green" are discussed without giving the specific conditions under which one might use them. One use of "This is not green" occurs when someone whose eyes are bad talks of a thing's being green, another, when someone is given a bag marked "green", meaning "Look in the bag". You must give the game in which the word or sentence is used, the circumstances under which you would use it.

Suppose I make up a new word "boo" and you bring me things until I say, "Yes, that is boo". Is it possible that you should recognize the color you have brought, which I call "boo" and which was never explained to you? What is the criterion for recognizing a color? Would you say you can recognize a face you have never seen before? No, for recognition of the face is actually taken as the *criterion* for having seen it. You cannot say that you must have seen it in order to recognize it, for that would be circular; recognition is the criterion for having seen it.

It all depends on the language game whether one says one must have seen green in order to say "This is not green". In certain games this is the case, in certain ones not. Suppose you say "I do not have pains in my hands". Most people think that if this is to make sense I must know what it is like to have pains in my hands. What is it like to know there are pains? Perhaps what is before my mind is some sort of shadow of a pain. The presence of the word "must" shows that there is something fishy here. It shows that nothing else will satisfy me. I could, for example, give myself a *sample* of pain, say by pinching myself. If I did not have this sample, if I had forgotten what pains were or how to produce them, would I have forgotten the meaning of the word?

What does it mean to forget the meaning of a word, e.g., "red"? One sense of forgetting what "red" means is ceasing to be able to imagine it again. But it does not necessarily mean being unable to recall an image, since you might be unable to form an image and yet be able to recognize red when you see it. *In some usages* "forgetting the meaning" might mean being unable to recall, in others ceasing to be able to recognize, without recalling. Both being unable to recall and being unable to recognize are things we call forgetting the meaning of a word or forgetting a certain use of a word. There is not one thing, but many, which we call forgetting the meaning of a word. To examine forgetting what a pain is like, look at its opposite, remembering what a pain is like. Remembering *may* be saying it was a terrible pain. It is sometimes not necessary to have an image (a shadow of the pain).

Lecture II

Let us return to the question whether a negation can be replaced by a disjunction. To answer this is to give the solution of a mathematical problem. It is important to remember that even if one were successful in replacing negation by disjunction, this solution would not help at all in getting on with philosophy. The solution of a mathematical problem *never* helps us in philosophy. Every mathematical problem is on the same level in this respect and is of no importance to us.

When one talks of the foundations of mathematics there are two different things one might mean. One might mean the kind of thing meant by saying that algebra is the foundation of calculus. In order to learn calculus one learns algebra. Mathematics in this sense is like a building, and in this sense a calculus such as *Principia Mathematica* is a bit of mathematics. The bottom layer is the one you begin with. One might also mean by foundations a means of shoring up something that is problematic. If there were something problematic about mathematics as such, then no foundation is less problematic, and giving one does not help. This is not to say that a calculus has no philosophical importance; it may be *very* important. The drudgery, the calculation, are unimportant, but the calculus may be useful philosophically in showing various things.

The introduction of Sheffer's stroke notation is a mathematical achievement. So likewise would be the replacement of a negation by a disjunction. The question whether negation can be "reduced" to disjunction has been put for an entirely wrong reason, and attempts to answer it have been made in an entirely wrong way. Whenever one asks whether the same thing can be expressed in a different way it is nearly always a mistake. For the question shows a wrong idea about symbolism. It is as though people thought the expression and what is expressed are in the relation of cause and effect, and that one could ask whether another cause would produce the same effect. This to draw an analogy where none exists. An expression and what is expressed are not in the relation of cause and effect. "These signs express this" misleads us thoroughly, as is shown by the fact that one gets to know what a sign means by learning its use, not by learning what effect it has on people. What it means is not a fact of natural history. I do not say that the effect is unimportant. The effect of chess, for example, is to give us entertainment, but this is not part of the definition of "chess".

How do we use "The sign '——' expresses . . ."? Consider the "The sign '~' expresses negation", which makes it sound as though the sign "~" were unsymmetrical, whereas it is symmetrical. This proposition is about the usage of the sign, i.e., that "~" = "not", and expresses a correspondence between two symbolisms.

To return now to the question, "Can negation be expressed by disjunction?" Suppose I said "Go anywhere, but not here", and that the order was replaced by one involving a disjunction. The disjunction would probably have the same effect as the original order. But what we want to know is whether it *expresses* the same thing. *It is something different,* if that is what you mean to ask. "Go anywhere, but not here" is not the same as "Go there or there or . . ." About the disjunction, one cannot say how many elements occur in it.

It is like a mathematical problem to ask whether one could find something equivalent to negation. It is *no* mathematical task to find a disjunction which could be substituted for a negation if there is no method of finding it. Here we have a case where to ask "Is there . . ." is to ask for the solution of a mathematical problem where we have no method. It is like asking whether a game could be constructed which one would be inclined to call a game with disjunction. Of course there really are cases where negation and disjunction are the same game, for example, the order "Write one of the permutations of *a, b, c,* but not *cba*". Within the grammar of our language we have a rule to replace this by a disjunction. But in the case of the order "Write down a cardinal number that is not 5", negation and disjunction are not the same game. We might be inclined to say it is equivalent to "Write down $1 \vee 2 \vee 3 \vee 4 \vee 6$ . . . etc." But "etc." is not a cardinal number, nor is $1 \vee 2 \vee 3 \vee 4$ . . . *ad inf.* a disjunction.

What is meant by an infinite disjunction? The phrase "infinite disjunction" is misleading because it suggests a huge disjunction. Suppose I replace "$\vee$ . . . ad inf." by "$\uplus$". This new symbol misses out something. It lacks the suggestion made by "$\vee$ . . . ad inf." that it is on the same level with "$a \vee b \vee c$ . . . $\vee z$". Of course there is a similarity between a finite disjunction like the latter and $1 \vee 2 \vee 3 \vee$ . . . *ad inf.*, but it is not in the italicized part, of which the distinguishing mark is not *ad inf.*, but $\vee$ . . . *ad inf.* The similarity lies in the italicized part of $1 \vee 2 \vee 3 \vee$ . . . ad inf. The real point of putting "$\uplus$" for "$\vee$ . . . ad inf." is that it shows up the difference between it and the finite case, whereas "$\vee$ . . . ad inf." does not. The "and so on" of an infinite logical sum is an entirely new sign with new rules. It does not correspond to any enumeration.

Suppose I said, Draw a circle in a square but not this:

Is this to be expressed as a disjunction of the same kind as $f(\sim 4) = f(1) \vee f(2) \vee f(3) \vee f(5) \vee$ . . . etc? No. There is this difference between "Draw a circle but not this one" and "Draw this or this or this . . .": the latter leaves open the possibility of your drawing the circle I do not want, but the number disjunction which leaves out 4 does not allow a similar possibility.

Questions about disjunction and negation are connected with questions about the different meanings of "all" and "any", the different kinds of generality, illustrated, for example, by "Draw *any* circle except this" and "Write *any* number except 4". Is the multiplicity of circles the multiplicity of what one calls real numbers? No, not if one is ordered to *draw* circles. We might come to a queer conclusion, that since only a finite number of circles is distinguishable we have here a finite disjunction. Now is this so? No. We do not even have a disjunction here, for there are no distinguishing marks in the language for the various circles. Similarly for the order "Paint me a shade between white and blue". There is not a finite disjunction here; there is not a disjunction. One feels like saying "One must mean *one of the possible ones* between white and blue, and one also feels that there are but a limited number of possible ones. But there is no means of naming them, and so a disjunction cannot be constructed. Suppose you mixed paint until you got blue, at each daub asking me if I could distinguish one from the other, and then asked me "Didn't you mean these?" What can this mean, that I *meant these?* One couldn't have meant them before they were given. These furnish one with a new bit of language. Of course I could say that it is a finite disjunction but one whose members I do not know. But it is in fact then not a disjunction.

Suppose I asked you to paint a circle inside a square, and that you did this:

Suppose I then argued: "Every circle in this square fulfills my order. This circle is in the square. Therefore this circle does." What sort of proposition is "This circle is in the square"? Consider in this connection Russell's reduction of "I met a man" to "There is an x such that I met $x \cdot x$ is a man". This way of writing generality did have the virtue of calling attention to the distinction between "I met a man" and "I met Smith", but in other ways it is enormously misleading. How are predicates used in our language? Russell uses "man" as a predicate, although we practically never use it as a predicate. (Just this sort of use often appears in philosophy.) Logicians use examples which no one would ever think of using in any other connection. Whoever says "Socrates is a man"? I am not criticizing this because it does not occur in practical life. What I am criticizing is the fact that logicians do not give these examples any life. We must invent a surrounding for our examples. We might use "man" as a predicate if we wanted to distinguish whether someone dressed as a woman was man or woman. We thus would have invented a surrounding for the word, a game in which its use is a move. It does not matter whether in practice the word has a place in a game, but what matters is that we have a game, that a life is given for it.

When "man" is used as a predicate, the subject is a proper name, the proper name of a man. I might give inanimate things proper names, though we usually do not. Suppose I have two exactly similar chairs and give them proper names, say "Jack" and "John". How shall I distinguish them? I must follow all their movements. It is assumed that the use of proper names is very elementary, but what are called proper names can be used in many ways which are not simple. What are the conditions for my being able to use "Jack" and "John" in the way we would normally be inclined to use names? One condition is that the two chairs could not be made to coalesce as do shadows, another that the path of each chair is continuous. This is a hint at the complications of the use of "proper name".

Lecture III

The term "man" when used as a predicate can be sensibly asserted, or sensibly denied, of certain things. It is an "external" property, and in this respect the predicate "red" is the same. But note the distinction between *red* and *man* as *properties*. A table can be the bearer of the property *red*, but the case with *man* is different. What is the *bearer* of this property?* The sentence "I see a man" is not explained by "$(\exists x)$ I see $x \cdot x$. is a man". For the latter leaves the use of x unexplained. It might be an explanation of saying "I see a man" if this were said of a dark patch in fog, or of a human-looking figure which behaved like a man, or of a roll of carpet with a man in it. Consider Russell's notation for "There is no man in this room": "$\sim(\exists x)x$ is a man in this room." This notation suggests one's having gone through the things in the room and found that none were men. The $(\exists x)fx$ notation is built on the model in which x is such a word as "box" or other *generic* name. The word "thing" is not a generic name. Suppose I translate "There is a painted box in this room", one obvious translation being one from which Russell's notation is taken, in which x is "box". Russell would not translate it in this way, but rather as "There is an x which is a box and is in this room". What is the x here?

Consider the notations $(\exists x)fx$, $\sim(\exists x)fx$, $(\exists x)\sim fx$, together with the example, "There is a patch in this square".

Put in Russell's way it would read: $(\exists x)x$ is in the square $\cdot x$ is a patch. What is it of which one says it is a patch? In contrast to $(\exists x)fx$, read in Russell's way, look at the notation $\sim(\exists x)fx$. This is sensible since it can be read as "There is no patch in the square". But consider

*Note how different are the cases of saying "This is a chair" to a blind man and to someone who visited the modern Finella House, where a piece of furniture might be different enough from conventionally designed chairs that one might well say this.

$(\exists x) \sim fx$. This notation is sensible if the x is interpreted as "a patch in the square", and "$\sim$f" as the predicate "not-red". But what would it be like for there to be an x which is *not a patch* in the square? Equally absurd is: "There is *not a thing* which is *not* a patch in the square".

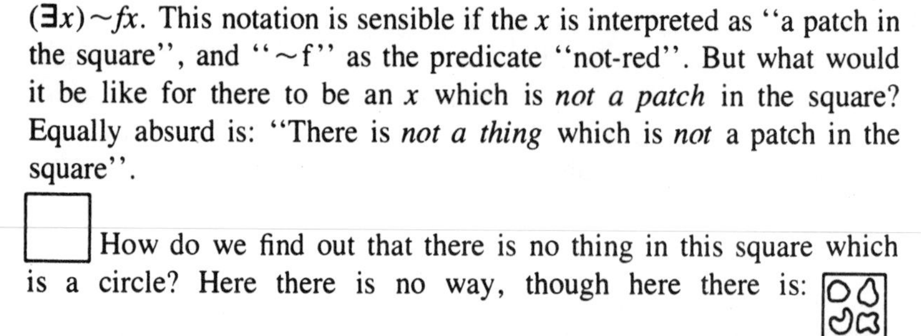

How do we find out that there is no thing in this square which is a circle? Here there is no way, though here there is:

For in the latter figure we have a case of something's being a circle or not being a circle. The proposition, "There is nothing in the square which is a circle", is utterly different in the two cases. In the second case it makes sense to say either that there is or is not a circle in the square.

The way a proposition is verified is part of its grammar. If I say all cardinal numbers have a certain property and all men in this room have hats, the grammar of each is seen to be different from the other because the ways of verification are so different. Moreover, to know that all men in the room have hats I must know not only the enumeration but also whether there are more than I have enumerated. The latter is an entirely different process, and there are many different ways of finding out whether the list is complete, almost as many as there are different cases. Again, note the difference between a hypothesis for explaining the motion of the earth, e.g., that a row of stars we see in the sky is much longer than our telescopes would reach, and the supposition that the row is infinite. Finally, compare "This square ☐ is all white" with "Every point in this square is white". The first does not mean the same as the second as is shown by the fact that it is not verified by going through every point. Compare the use of "every point" in the translation of "This square is white" into "Every point is white" with its use in the translation of "finding the maximum of a function" into "finding the point that is higher than every other". Since points are not things that can be seen and singled out, like patches, "Every point is white" is misleading. This translation suggests that we have an explanation of an ordinary proposition, a more scientific analysis. If the square contained tiny circles painted white so that the whole square appeared white, this explanation of its appearing white would be an analysis. Without some such explanation the translation into "Every point is white" is all right only if you see it means not one jot more than "This square is white". In the case of finding the maximum of the function $\frown$, we seem to have found the highest point; but again, we cannot look at every point to ascertain this. We use a trick to verify the proposition that this point is higher than any other point of the curve. And thus the proposition has a different grammar from a proposition verified differently.

Russell thought to describe the foundations of arithmetic by giving a theory of propositions, functions, etc. He treated propositions and functions as a uniform class. I seem to be saying that every proposition is different, and thus I may seem to be denying the generality of mathematics and arithmetic. If we wish to begin as Russell did, and if "proposition", "function", "generality" mean all sorts of things, then it looks as if we shall have many arithmetics, different arithmetics for patches, men, thunderstorms, committee meetings, etc. We might ask then how it can be that the arithmetic we learned at school holds good. The generality of arithmetic is not threatened in the least. To understand this we must look at the way arithmetic is *applied*. What is the relation between mathematics and its application? Arithmetic is a calculus, and is in roughly the same relation to its application as a paradigm is to what it is a paradigm of. In elementary school we learn arithmetic by counting beads on an abacus, or by using physical objects for illustration. Later we calculate with numbers without reference to any particular objects. But this is not because arithmetic is "general". Arithmetic is like an instrument box—like a box of joiner's tools—and we can be taught the use of the instruments. But though their uses are explained, one would not have obviated the need to explain how to make a chair or table, or to deal with various woods. Each of these requires a slightly different use of the instrument, just as every wood requires different treatment. The explanation of the use of an instrument is a preparation. Teaching will give certain rules for its use. Then we will see what it can be applied to.

Arithmetic is not taught in the Russellian fashion, and this is no inaccuracy. We do not begin arithmetic by learning about propositions, and functions, nor with the definition of number. And this is not because children cannot understand these things. The way we learn arithmetic is the proper way.

Lecture IV

The use of proper names is regarded by logicians as something very simple and straightforward, but it involves complicated questions about identity and continuity. How do we use the subject in the following sentence: "*This* is now circular but half an hour ago *it* was not"? Patches can, of course, be given names. Of *A*, where "*A*" is the name given *this*, we can either say *A* has changed in the half-hour lapse, or that *A* is a new patch, or if change has been continuous, that an infinite series of patches has come into existence. And if the patch vanishes every now and then, we can lay it down as a rule that it is the same patch if it has the same size. Or if there are two where there was one before, we can say *it* has split up. If two moving shadowy patches, given the names *A* and *B*, coalesce at the intersections of their paths

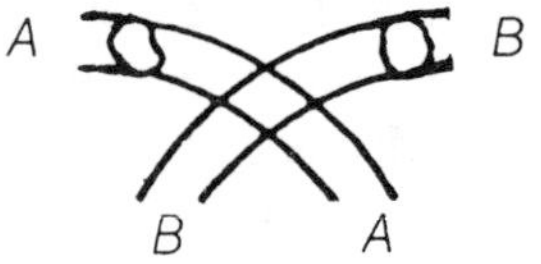

A B

B A

we can lay it down that the patches at the corners are called *A* and *B* as in the diagram, or we can lay down just the reverse. We can make any rules we please. Suppose we had a ring of patches of different sizes, , and that they changed sizes suddenly so that one became the same size as the next all the way around. Under certain circumstances we would say the patches had moved, under others that the same patches had shrunk or grown, as the case might be. Similarly, one could describe the light of the electric sign  either as moving back and forth, or as swelling and diminishing.

There is a fundamental confusion about questions regarding sense data, the confusing of questions of grammar with those of natural science. For example, Is whiteness circular or is a patch white and circular? What makes the question attractive is that the answers appear to decide between *existence* and *nonexistence* of something, namely a *patch*. We are really just turning our language around here when we ask "Is there a patch or not"? For "Whiteness is circular" and "The patch is white and circular" say the same thing. The philosopher does not tell us how to decide the question. The same is true of the question whether or not a sense datum is identical with, or part of, the surface of an object, and of the question whether the chair or its surface is brown. If these were questions of natural science we should need to be told how to decide them, what the method of verification is. The question whether a body seen through a glass is yellow or the glass yellow is a sensible one, for one knows how to find out which answer is true. But how to decide whether whiteness or a surface or a sense datum is circular? The philosopher does not tell us how to decide between these, and what is more confusing is that often a question such as these *has* an application, which makes a philosopher think it has when he asks it.

Usually a difference of opinion in a certain situation is indicated when one person says this and another that. But a difference of opinion does not always show up in this way, and the fact that two people say different things is not always a sign of a difference of opinion. The man who says "Here is whiteness which is round" and the man who says "Here is a patch which is white and round" say the same thing. Similarly when one person says that a surface has changed, and another person says it is the same surface which appears different to him. There would only be a difference of opinion if the two statements "The patch is circular" and "Whiteness is circular" belonged to the same game. These expressions stand in different calculi, and different things are done with them. Belonging as they do to different games, they can appear to express different things while expressing the same thing. If two people forgot that they had different systems, for example, one in which a fair-haired person is painted with fair hair, and the other in which nothing is painted on his head, then it might happen that the two would ask whether

has fair hair or none. Similarly if one had a notation, "There is a table", for what is ordinarily expressed by "There is *one* table", a confused question could arise: "Is there a number of tables or not?" And to the question, "Isn't the latter notation more *adequate,* more *direct?*", I would answer No. For one symbolism does not come nearer the truth than the other. (Of course it is all right to ask whether one symbolism is more misleading than another.)

What I want to say is that these questions are treated fundamentally wrongly. To ask, "Are there only sense data, or physical objects as well?", sounds like "Are there electrons or can we manage with only protons?". The two are entirely different.

Lecture V

Suppose one asked whether a drawn square is complex or simple, i.e., whether it consists of parts or not. One might reply, "I could draw a line dividing it into parts". But what if it is too small to bisect? We could lay it down that it has parts if one could draw a dividing line or bisect it as the mathematician does. The rejoinder to the question, "What if it is too small to do either?", viz., that I can imagine a dividing line, is very peculiar. For we say this whether we can or not. The reply does not mean just what it seems to: merely that it can be imagined. At least we don't make a picture by imagery; we make a picture by description. It is this which is so important, for by giving the description we imply that it makes sense to say it is divisible. (Compare the series of fractions, each member of which "can be divided".)

Suppose I have a very accurate dividing machine, and that I say the square is divisible if I can divide it with this machine, otherwise not. "Can" here stands for a physical possibility. "Can" in "can imagine" refers to the possibility of imagery. Having agreed on the use of the word "divisible" we can say whether the square is complex or simple. We shall call it complex if it is divided into patches we can see. The question whether it is complex can be answered by a factual statement. But besides this question there is the *philosophical* question, using the same words, namely, "Is this uniform white object complex or simple?" The answer is, "It depends". Here we find ourselves unravelling a philosophical problem. When a real question corresponds to the philosophical question, as here, it is easy to correct the mistake, but sometimes it is very difficult. It seems as if there must be *one* answer to all such questions.

Let us return to the example, "There is a cardinal number which I can write down", of the form $(\exists x)fx$. This is not a disjunction since "and so on" is no numeral. But one is tempted to say it must be a disjunction of the form $fa \vee fb \vee fc \vee \ldots$ because one can infer $(\exists x)fx$ from fa or from fb or from $fa \vee fb$. What tempts one is that $(\exists x)fx . \vee . fa : \equiv . (\exists x)fx$ and that $(\exists x)fx . fa . \equiv . fa$. For seemingly the addition of the disjunction adds nothing, which *could* only be if $(\exists x)fx$ is already a disjunction—already contains a disjunction. Just as $fa \vee fb \vee fc . \vee . fb : \equiv . fa \vee fb \vee fc$ because fb is already included, so one is tempted to say that $(\exists x)fx . \vee . fa : \equiv . (\exists x)fx$ for the same reason.

The entailments $fa. \supset .(\exists x)fx$ and $(x)fx. \supset .fa$ seem simple when $(\exists x)fx$ is viewed as a logical sum and $(x)fx$ as a logical product.

An analogous question arises when one proposition p follows from another proposition q: Mustn't one think p when one thinks q? In general, mustn't one think the conclusion in thinking the premise? This is important because there are several wrong answers. The affirmative answer has a deep reason back of it.

Are the steps taken in following a rule contained in it? Suppose someone is given the order, "Write down a cardinal number", and he fulfills it by writing 127. Was he not ordered to write this number together with other alternatives? Suppose he is taught to write an arithmetic series by adding 1 to a number, 1 to that, and so on. The teacher trains him by examples to carry out orders in accordance with the rule "Add 1". Suppose now that he is ordered to add 10, and that the highest number reached in the training is 100. Upon being given the order he writes 10, 20, . . . 100, 120, 140, 160, and the teacher objects that he did not carry out the order. But why? The teacher replies that he was meant to do this: 100, 110, . . . 1,000,000, 1,000,010. and when did the teacher mean it? When he trained him. And *up to when* did he mean it? It is strange that he should have had time to mean all this. The assertion that the teacher meant this when he trained him is terribly misleading. For it suggests that another process was going on during the teaching, or that even though there was no process going on corresponding to each step, there was a process going on containing all these steps and from which these followed, a queer process containing all these unborn steps. If what the teacher meant did not contain all these steps, how explain his knowing at once that the pupil was wrong?

However, the statement, "I did not mean you to write 120 after 100", is not really an account of what the teacher did earlier, but what he is doing now. He might justify himself by "I would have told him that 110 was the next numeral after 100 if I had been asked". This is either a hypothetical statement, or a rule. But *this rule was not given.* It does not alter matters to say the teacher would have told him this, even though he were believed.

Now is the person wrong who writes 100, 120, . . . ? Couldn't he even show you he is right? He could cite a rule which so far as his training went was the same as the rule he applied: Add 10 up to 100, 20 up to 200, etc. But apart from this, "Add 10" could have been so used that his next number after 100 was 120. It has not been made impossible to do this if his following the rule extends past 100 and what he was to do after 100 was not mentioned in the training. It is even possible that after 120 he stops. The pupil is given a rule and examples, and the teacher may say that he *means* something, that though not stated is conveyed indirectly by means of these. It would seem that if what is meant could be conveyed, and not merely the clumsy rule and examples, he could be *made* to continue with 110 after 100. But the teacher also has only the rule and examples. It is a delusion to think that you are producing the meaning in someone's mind by indirect means, through the rule and examples.

Lecture VI

In saying it is understood that the person who follows the rule "Add 10" is to add 10 to each number, we refer to the way he is taught. If he surprises us, after being taught addition up to 100, by continuing with 120, 140, etc. and we say "I *did* not mean this", our use of the past tense creates the delusion that something else happened at the time of training than actually did. This delusion is but a special case of another one: that the chain of reasons has no end. Why *must* one write 110 after 100? Is there an answer to this question? There is, namely that this is what one usually does after instruction. But isn't there another answer? Couldn't we answer the question "Why did you add 10 when given the rule?" by giving another rule for following the rule "Add 10"? A reason *need* not be given in answer to this question, but one *can* be. Suppose we had the chart

a	A
b	B
c	C

and that we were trained to translate by means of it any small letters such as *aabbc* into capital letters. The chart justifies the translation into *AABBC*. Now if one is asked for a reason or justification for using the chart to translate in this way, one could give the schema

to explain the rule given by the chart. And for this schema you *could* give another schema justifying it. The chain of reasons *may* end with the chart, but it need not. When a person translates by the chart alone, without being given the schema of arrows, did he know this latter rule? It might be argued that if he had not known it he could not have used the chart as he did. This makes it appear that the chain of reasons has no end, that only the reasons written down have an end. But *must* one know this rule in using the chart? No. One simply makes the translation. The answer to the question, "Why, having been trained to translate by means of the chart, did he write *AABBC?*", is merely that he did it—unless one cites another rule. It might be objected that if one just does it this way one acts like an automaton, without understanding. But in understanding something one often just *does* it.

This example is precisely like the example of the rule "Add 10", which may or may not have a rule back of it. To say that if one did anything other than write 110 after 100 one would not be following the rule is itself a rule. It is to say "This rule demands that one write 110". And this is a rule for the application of the general rule in the particular case. Note that this rule was not given in the training, unless by accident, in which case there would be other rules that would not have been given.

We can say that the order "Write 110 after 100" follows from the rule "Add 10". This is what leads us to say that in giving the rule "Add 10" we meant that 110 follows 100 (to be symbolized by $100 \rightarrow 110$), and thus that $100 \rightarrow 110$ follows from the rule. We could then say that $100 \rightarrow 120$ contradicts the rule; and there is the temptation to say that it could not do so unless $100 \rightarrow 110$ were presupposed in it. In what sense is it preformed, or presupposed, in it? If in following the general rule a person said he must write 110, then the question, Was he making a new discovery?, is puzzling. The form of the question suggests that he was, that it is like a scientific discovery; yet to say that he has not made a new discovery suggests that $100 \rightarrow 110$ is presupposed in the rule "Add 10". There is actually no new discovery here, although $100 \rightarrow 110$ is not contained in the rule. When he said he must write 110 after 100 in following the general rule, there is no question of *discovering* a step the rule compels; rather it is a question of a new *decision*. The decision, unless made in the training by accident, has not been made. In the case of translating our chart, a new decision is made at each use of it.

Mathematical intuitionists have said that one needs a new intuition for each step taken, say, in developing a progression. What they saw was that giving the general rule does not compel one to make the step. It is wrong to think one takes the step by insight, as if one no longer has any reason but a sort of revelation instead. In saying there is a process of intuition it seems to be explained why one could be so clever as to write 51 after 50! If any mental process is involved, it is one of decision, not of intuition. We do as a matter of fact all make the same decision, but we need not suppose we all have the same "fundamental intuition".

It will help greatly if one once gets the idea of a philosophical delusion.

To return now to the question: Is the conclusion thought when the premise is? If the word "thought" is replaced by the word "said" in this question, the answer is No. But there is a deep meaning in the question whether when the process of thinking the premise went on, another process, of thinking the conclusion, went on. If it is claimed that the question really asks whether the fact that a certain conclusion follows from a given premise is a discovery, the answer is No, it is no new discovery. This answer tempts us to reply, "Then the conclusion is thought when the premise is". *In the sense* that a certain conclusion's following from the premise is no new discovery, no new phenomenon, it is correct to say this. But if you try to make out that this special rule, which has never been given, has in some way or other been given with the general rule, this is nonsense.

Suppose we have a curve with two tangents at the points A and B

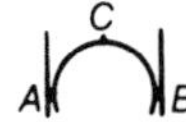

and that we roll a ruler around this curve from A to B. Is it correct to say it must have passed C since it must have passed *all the positions* between the two? A simpler example which raises the same sort of question is the sliding of a ruler from 0 to 4 along this line: 0 1 2 3 4. To say that the ruler *must* have passed through all the rational and real numbers from 0 to 4 makes one think that though one cannot calculate all the real numbers, for example, π, one has actually been exactly at π! Since I did not know I was at π, does not this seem a wonderful achievement! Now was one at the point π? One answer is: We pass through 3 and π in different senses: we see the ruler slide over 3, and we've got to lay it down that if it passes from 3 to 4 it passes π.

What does one call passing from 0 to 4? Why are we tempted to say that in passing from 0 to 4 we must have been at the point midways? Is there good reason for saying this? *Were* we at all the points between 0 and 4? If by moving from 0 to 4 is meant passing the points one sees, and by "all points", more than the visual points, then we did not visit all the points. It is a different thing to use the words "We passed from 0 to 4" to refer to a visual phenomenon, and to use the same words to make a rule about the way we are to talk.

In sliding the ruler from 0 to 4 could one miss out a point? In what sense is it the case that one might have? If the movement of a hand, or shadow, is visually continuous, must the hand have passed every point on the scale? Suppose it vanished for $\frac{1}{1000}$th of a second. Was the movement then continuous or discontinuous? Did it pass through all the points or not? The answer is "It depends". Visually it is continuous, but physically it was discontinuous. And if by not missing out any point, or by passing through all points, you mean it was visually continuous, then it did not miss out any point; it passed all points.

Lecture VII

What is the criterion for a proposition being a proposition of logic? One claimed criterion is self-evidence, which seems to be a psychological criterion; and yet the self-evidence seems to be in the symbols, "objective".

Frege had the idea that every sign, proposition as well as descriptive phrase, had a sense and a meaning.* Two signs might have the same meaning but different senses. And he went on to say that a proposition has one of two meanings, the true and the false. "*p*" means the true if "$\sim p$" means the false. The function $\sim p$ was treated as a coordination of the two values, the true and the false, and a table could be written for it:

p	$f(p) = \sim p$
T	F
F	T

Frege did not see that this table can itself be taken as a symbol for the function, though it looks as though it says something *about* the function. Frege has instead only given a translation: "$\sim p$" translates as

p	
T	F
F	T $= \sim p$.

This schema does not say anything about $\sim p$; it is another way of writing it.

Frege explained the notions of "or" and "not" by the notions of

*It is more usual to translate Frege's distinction between *Sinn* and *Bedeutung* as "sense" and "reference". (Editor)

"true" and "false". That $p \vee q$ is false only if both p and q are false states a rule, and is embodied in the truth-function symbolism as follows:

p	q	
T	T	T
F	T	T
T	F	T
F	F	F

Written as a row: (p, q) [TTTF]*

It is important to see that this table, like the table for $\sim p$, says nothing about $p \vee q$, but is another way of writing it. When Frege explained such functions by listing the truth-values of the arguments in columns on one side and the function on the other, it looked as though he had said something *about* the function. But instead he had defined it, given another notation for it.

My object [in the *Tractatus*] was to show the essential difference between a symbol for a proposition and a descriptive phrase. A proposition p was written with two poles†, TpF, and the combination of truth-possibilities of p and q with lines as follows:

$p \supset q$ is written:†

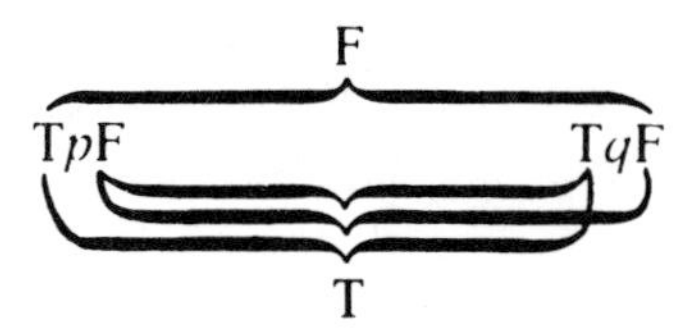

Now if you look at $p \supset p$, you can see what characterizes a logical proposition. Here we have one argument p, which has the two truth-possibilities T and F. The proposition $p \supset p$, has only one pole, the true pole:

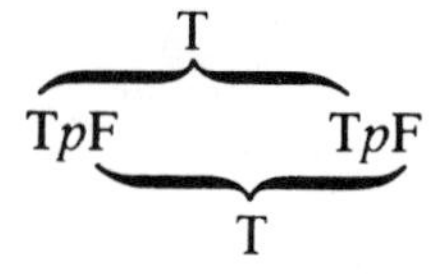

*See *Tractatus Logico-Philosophicus*, London and New York, 1922, 4.442. (Editor)

†See *Tractatus* 6.1203. (Editor).

Written in accordance with this simple rule a logical proposition is distinguished from any other. The important point which the schema shows is that logical propositions have nothing to do with self-evidence.

Let us look at what sort of proposition $p \supset p$ is. The table makes it clear what we mean by saying propositions like it are redundant. There are two puzzles: (1) Why, if they are tautologies, do we ever write them down? What is their use? (2) What sort of generality has the statement "This is true for all propositions"? (e.g., that they cannot be both true and false). Further, if all tautologies say nothing* then don't they all mean the same? $p \supset p . \equiv . p \vee \sim p . \equiv . \sim (p . \sim p)$! All have the same sense, viz., no sense! Difficulty is not only posed by the question, "What are they used for if they have no sense?", but also by the question, "Why do we use so many of them if they have the same sense?" Let us examine the use we do actually make of them. First, we do not inform by means of them. When in all the blanks in the truth-table for a proposition there appears a T, it does not even seem judicious to call it a proposition. But to call it nonsense is also not judicious, because it is unlike "yellow tables chairs" and "the slithy toves gimble", which we call nonsense. When I called tautologies "senseless"† I meant to stress a connection with a quantity of sense, namely 0. "$p \supset p$ is senseless" could be translated as "$p \supset p . q . \equiv . q$", to show up its continuity with other propositions, and its difference from them. A tautology is a degenerate case of a proposition. It plays something of the same role in logic as the 0 of arithmetic. $0 + 2 = 2$. Analogously, $p \supset p . q$ is like $0 . q$.

Now what is the use of all these propositions? Let us examine one which has played a role in logic: $p \supset q . p . \supset . q$. Here we have a tautology, as shown by its truth-table, although it seems to say something, inasmuch as we make inferences in accordance with it. By itself it is not a rule of inference, for a rule should say something, and $p \supset q . p . \supset . q$ says nothing. That it seems to say something is because the second sign of implication seems to say something the first does not, something having to do with the word "follows". Inferring is connected with the second sign, not the first. What is the connection between implication and inference, between $p \supset q . p . \supset . q$ and the in-

* See *Tractatus* 4.461, 5.142, 6.11 (Editor).

† *Tractatus* 4.461. (Editor)

ference $p \supset q$? Does one need the former as a link in this inference?

$$\frac{p}{q}$$

And if so, then why not something to connect its use as a link? The proposition $p \supset q . p . \supset . q$ is a pattern by means of which the conclusion q is inferred; but q is not inferred. What allows the inference of q is not what the proposition *says* but the fact that it is a tautology. The rule of inference is not $p \supset q . p . \supset . q$ but "$p \supset q . p . \supset . q$ is a tautology". The use of this rule is to make an inference from one *ordinary* proposition to another. Such a rule is to be distinguished from a rule of inference in a logical system. A rule of the latter kind is applied to the primitive propositions and their consequences, not to ordinary propositions.

Lecture VIII

There were important reasons for puzzlement about the general propositions $(x)fx$ and $(\exists x)fx$. One was that there must be something, so I thought, that can be called the general form of a proposition, something common to all propositions. Another was that if $fa \supset (\exists x)fx$ asserts the relation of *following,* then it must have something in common with the tautology $p . \supset . p \vee q$. $(\exists x)fx$ ought to contain fa as one truth-function contains another. I had the mistaken idea that propositions belong to just one calculus. There seemed to be *one* fundamental calculus, viz., logic, on which any other calculus could be based. This is the idea which Russell and Frege had, that logic was the foundation of mathematics. The task was to exhibit what is characteristic of this one fundamental calculus, to show what logic is. Logic treats of propositions and functions, and mathematics could be based on logic. Thus logic gives the general form of mathematical propositions. It seemed to me that the words "proposition", "sense", "generality", "logic" were all equivalent to each other. If one has the idea of a single logic then one must be able to give one general formula of logic, the general formula of a proposition. I thought I had found this formula in the T-F table, an equivalent of the word "proposition" and the word "logic".

The idea that logic gives the general form of a mathematical statement breaks down when one sees there is no such thing as one idea of a proposition, or of logic. One calls lots of things propositions. If one sees this, then one can discard the idea Russell and Frege had that logic is a science of certain objects—propositions, functions, the logical constants—and that logic is like a natural science such as zoology and talks about these objects as zoology talks of animals. Like a natural science, it could supposedly discover certain relations. For example, Keynes claimed to discover a probability relation which was like implication, yet not quite implication. But logic is a calculus, not a natural science, and in it one can make inventions but not discoveries.

I am not taking the view of C. I. Lewis and the Warsaw schools that there are many different logics. In speaking of more than one logic I am not referring to non-Aristotelian logics such as the three-valued logic in which propositions had three possibilities instead of two, T, F, and Possible. There is great danger in making up such a game, unless taken as a game. The value of such games is that they destroy prejudices; they show that "it need not always be this way". But if this latter is said as though it were a statement of science (like "You think all rats are like this, but there are others"), then the 3-valued system, for example, might appear to be an *extension* of logic, representing a discovery.

Now who uses the calculus of T and F? I would say it has no use. Taken as a calculus, it is dull and useless, and so is Russell's calculus. But it has a justification which may not hold for other logics. The point of the T-F calculus is to afford a translation of Russell's calculus, making clear the relations between the latter sort of calculus and its application. A calculus is of no value unless it makes one clearer about another.

Lecture IX

What is the connection between the idea of a proposition and the laws of logic? There is a temptation to say that $\sim(p\,.\sim p)$ and $p \vee \sim p$ are laws about propositions. Being tautologies, they say nothing; so in what sense can one say these hold for all propositions? And how does one know this? One has not examined all propositions. Our way of talking about them is misleading, in suggesting that we are saying something similar to "All apples are sweet". The latter proposition is different; here we have an hypothesis, which if true must hold for this particular apple if it holds for all. To say that these laws hold for all propositions seems to permit saying the same sort of thing: that in holding for all they must hold for this particular instance.

Why does one look at these two laws as fundamental? Because we see the similarity of $\sim(p\,.\sim p)$ and $p \vee \sim p$ to true propositions we make the mistake of saying they are true. It is important to note that we talk of the law of excluded *middle*, where no third alternative is ever mentioned. What is the middle which is excluded? $p \vee \sim p$ is modelled after "This is either red or green; there is no third alternative". But the comparison is a false one. This proposition has a very impressive form, and it is universal.

Brouwer talks of a range of propositions for which the law of excluded middle does not hold; in this branch of mathematics this law does not *apply*. Now note that to say a proposition is that which is true or false is to say it is that for which the Aristotelian laws of logic are valid. This in a sense gives a definition of "proposition". $p \vee \sim p$ and $\sim(p\,.\sim p)$ are rules, rules which tell us what a proposition is. If a logic is made up in which the law of excluded middle does not hold, there is no reason for calling the substituted expressions propositions. Brouwer has actually discovered something which it is misleading to call a proposition. He has not discovered a proposition, but something having the appearance of a proposition. The situation here is similar to that of a board game which is more analogous to a tug-of-war than to chess, yet keeps the appearance of being chess. The way in which mathematicians express themselves is taken from the language of natural science. To say the law of excluded middle does not hold for propositions about infinite classes is like saying "In this stratum of atmosphere Boyle's law does not hold".*

The definition of a proposition as anything that can be true or false, and thus as anything that can be *denied*, sounds as if it gave a criterion for deciding whether something is a proposition: deny p and see whether the result is a proposition. What is it like to try to deny p and to decide that it is no proposition if one cannot? The words "true" and "false" sound like the adjectives "red" and "green", and to say a proposition is whatever can be true or false sounds like a discovery, like "iron is anything that can rust". But to say a proposition is what can be true or false comes to saying that what we mean by "proposition" is partly given by the rule $p \vee \sim p = \text{Taut}$. It is what this and other rules apply to. And this means that these rules determine the game played with "proposition".

Suppose it is asked whether this game can be played only with

*This paragraph is for the most part taken from The Yellow Book, as a supplement to the material on the law of excluded middle given in this lecture. (Editor)

"proposition". Could it not be played with a word like "table"? In answer to this, let us examine two versions of the law $p \vee \sim p$, "A proposition is what is either true or false", "A proposition is what *can be* true or false". The word "can" in the second version introduces a dangerous element. To say that one can negate only a proposition presupposes the possibility of a kind of trying. It suggests that we have negation here and a proposition there, and that negation is the kind of of thing that will fit a proposition and nothing else, like a shape being fitted to certain other shapes. Suppose someone said, "I can negate 'apple'—I need only say that 'apple' is not true". The reply would be that "negation" and "apple" don't fit. If the rejoinder were, " 'Apple' is not true" means "Apple is not sweet", then the natural reply, namely, that the meaning of "negation" has been altered, creates the illusion that we have two ideas, "not" and "apple", which do not fit, as if there were a fitting or not fitting of two games that we play with "not" and "apple". The notion of a fitting is wrong. To negate an apple sounds like doing something with it, like eating it, whereas all that is done is to write two separate scratches, "not" and "apple", side by side. What corresponds to our idea of negation is the use we make of it. If for the variables p and q of Russell's calculus we substituted *words* like "apple", the result is something of which we do not make any use. Perhaps a use could be made for it. In fact we do have a game in which "not an apple" is used when we are refusing an apple. We might say that the gesture of refusal or pushing away is the meaning of "negation". This is the kind of use of "negation" which occurs in the primitive language between the builder and his mate. "Not brick" would be translated into our language as "don't bring me a brick", and "Apple" into "Bring me an apple". But not in the primitive language. In the game in which one says "Not an apple" there is nothing which one could call a proposition. There are two possible answers to the question why we should be tempted to call "Apple", or "Not apple", propositions, (1) that these are *the* sample propositions, which we could not say, and (2) that we wish to show a family of transitions to these special cases. To the question whether $p \vee \sim p$ can be used in a game where "apple" is substituted for p, my reply is that perhaps a game—a use—can be found, i.e., if a man were so trained that "apple" functions to produce the reaction which "Bring me an apple" would, that is, that the practice is the same. Note that the practice does not enter into the language; it is not given by the rules of the language.

A distinction is to be made between *use* and *application*. When I talked of "Apple" being used as an order, we understood this because we have here a practical application, which is useful. Whether an application has a use in practice depends on the kind of life we lead. The pragmatic criterion of the truth of a proposition is its usefulness in practice. But the person who says this has in mind one particular use of "useful": its use in the lab, say, to predict the future. But if a mad physicist were to offer a prize for a completely wrong hypothesis, then a person whose hypothesis had a distribution of confirmations like this 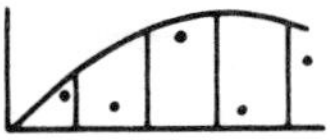 instead of along the curve, would find it useful although it was useless for prediction.

In contrast to the traditional logic, Russell introduced symbols for relations of two or more terms, with the idea of building up a logic to apply to all eventualities. Now what sort of proposition is "Love is a 2-termed relation? Obviously one has not said anything about love. Suppose someone doubted whether there are 2-termed relations, and someone replies, "I've found one, love". Has he made a discovery? It sounds as though he had found a natural phenomenon that fitted this schema, and that without it the schema would have been empty. If this is the case, then it would seem that the use of expressions for 3-, 4-, and 5-termed relations depends on natural facts. But we must remember that two people not being in love is also a 2-termed relation. If I came of a tribe where love was unknown and went to another tribe where I found someone in love, would I have discovered a 2-termed relation? One could say, "I have discovered a use for the word 'love' ". Here in this symbolism I have a use for "love" and need not look further. We can discover a use for symbolism in the sense of finding it useful (like the physicist and his wrong hypotheses), but there is no such thing as discovering a use for it in the sense of discovering a natural phenomenon which gives content to it where before it was empty. It is absurd to look at a 13-termed relation as empty until we have found a 13-termed phenomenon, for the calculus we make with these words does not receive any *content* from what is found; it remains a calculus.

Russell thought that in treating foundations he had to arrange for the *application* of arithmetic, for example, to functions. One could not talk about 3 apart from some type of function, so one would need to classify functions. Number is a property of a function. Russell and F. P. Ramsey thought that one could in some sense prepare logic for the possible existence of certain entities, that one could construct a system for welcoming the results of analysis. Beginning with 2- and 3-termed relations, of which one has instances, one could claim to have prepared the calculus for 37-termed relations, of which there are no instances. We tend to think that when we have found an example of *aRb* we have found a phenomenon to which *aRb* is applicable. We have only found a word in our language which behaves like *aRb*. Before an instance of *aRb* is found there could be the word in the language. Constructing a relation does not depend on finding a phenomenon. Discovering a word game is different from discovering a fact.*

In language as we use it there are not only words and their combinations but also words which make reference to *samples*. The word "blue", for example, is correlated with a certain colored patch which is a sample. Samples such as this are part of our language; the patch is not one of the applications of the word "blue". The phenomenon of love plays the same role as the patch in the use of the word "love". Two people in love may serve as a sample, or paradigm. We might say that it is the paradigm which has given the word "love" content. But for this purpose we need not discover two people in love, but rather the paradigm, which belongs to the language. We can say the paradigm gives the word meaning. But in what sense? In the sense of *enlarging* the game. By bringing in a paradigm we have altered the game. We have not found a phenomenon which gives the word sense; we have made up a calculus. To say that the paradigm fits the symbol, e.g., that the blue patch fits the word "blue", means nothing. It is *added to it*. And the schema is now useful.

The attempt to build up a logic to cover all eventualities, e.g., Carnap's construction of a system of relations while leaving it open whether anything fits it so as to give it content, is an important absurdity. We must remember that if we feel the need of an instance of an *n*-termed relation we still have the symbolism for *n* things not standing in relation. The need is for a *sample,* a paradigm, which is again *part of the language,* not part of the application. Samples play the role played by the Greenwich foot, the existence of which does not prove that anything is a foot long. The Greenwich foot itself is not a foot long. To say "Here is an instance of people being in love" is to take a sample into our language. And to do this is to make a decision, not to discover anything.†

*This paragraph is taken from the 1932–33 lectures entitled "Philosophy". (Editor)

†For discussion of this topic, see *Philosophische Grammatik,* pp. 309–14. (Editor)

Lecture X

Let us consider negation further, in particular, negation of "apple". What is the criterion for having negated it? If the criterion is merely writing "not" before "apple", then we have done it. If the criterion is that the combination of signs be useful, then obviously it can be made useful. If it is required that the phrase be accompanied by some feeling or gesture, then why should this not happen? None of these criteria is satisfactory. We want "not" to be *used* in a certain way. You were uncomfortable about my use of "not" with "apple"; but this could not have been because we do not use it thus, inasmuch as we sometimes do. What you must mean is that you do not want to use it in that way. You want to say that the use of the word "not" does not fit the use of the word "apple".

The difficulty is that we are wavering between two different aspects: (1) that *apple* is one thing or idea which is comparable to a definite shape, whether or not it is prefaced by negation, and that negation is like another shape which may or may not fit it:

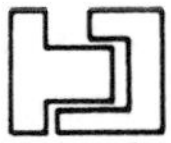

(2) that these words are characterized by their use, and that negation is not completed until its use with "apple" is completed. We cannot ask whether the uses of these two words fit, for their use is given only when the use of the whole phrase "not apple" is given. For the use they have *they have together*. The two ideas between which we are wavering are two ideas about meaning, (1) that a meaning is somehow present while the words are uttered, (2) that a meaning is not present but is defined by the use of the sign. If the meanings of "not" and "apple" are what is present when the words are uttered, we can ask if the meanings of these two words fit; and that will be a matter of experience. But if negation is to be defined by its use, it makes no sense to ask whether "not" fits "apple"; the idea of fitting must vanish. For the use it has is its use in the combination.

When we say it is impossible to negate a thing, that only a proposition can be negated, it seems (1) that it is not an experiential statement, but (2) that we can describe *what we cannot do*. If, however, one can describe this, then one can, except for human frailty, do it. What cannot be described because it is forbidden by the rules, one cannot do. But now by saying that we cannot use "not apple" as we can "$\sim$(apples are red)" we have not settled how we are to know when to use "$\sim$" and "apple". Does the use of "apple" itself exclude negation? We fix the grammar of "apple" and "not" by considering the rest of their uses. But the question remains how we are to use them in a case where we have not used them. What we must do is to lay down rules, and then it no longer would be a question of the word "not" applying or not applying to "apple" but of there being a use fixed beforehand.

Fixing the use of the sign of negation means settling what is to be put inside the brackets: ~(   ). If we say we do not want to substitute the name of a fruit, then we have fixed that much. Suppose I drew a circle and said there can be apples anywhere in the universe but not in it. Could not this mean "not apple"? That will depend on how the use is fixed. But we are not fitting together separate things like separate solid bodies.

Does it determine the grammar of the word "not" to say it can be used in connection with the word "apple"? Suppose that instead of using the word "not" I use the word "to", and that I said "to" fits "apple". What do you know? I have told you nothing at all, for you do not know in what way it "fits" "apple". I have only said that "to apple" can be written in this game. If "to" means the same as "sour", then it fits. In order to explain in what way it fits "apple", I should have to explain the way the *combination* is used. The question whether "to" fits "apple" suggests that "apple" already has a grammar, and to tell what fits it is to tell what "to" is like. But this is not the case. If one says "apple" means what it usually does, you can reply: "Then 'to' says something about 'apple', but I do not know what". In one sense I could say I have given you some information about "to" fitting "apple" if I said "to" can be used with "apple" just as "not" may be. This comes to saying that "to" = "not". To change the example, suppose I said "go is red". You will say that "go" must be the name of a particular or generic spatial object, or an afterimage. But "go is red" might be "nothing is red". In a sense it does give one some information about the use of "go" to say it fits "is red". But one cannot lay it down that the use of a word is given by telling what words fit it. Even ostensive definitions do not fix the use.

The statement "This fits that", asserted of two bodies

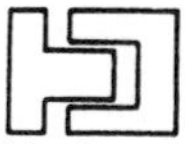

may be either of two different kinds, a geometrical one or an experiential one. If the diameter of the tongue of the left-hand piece is 3 inches and of the corresponding depression of the other piece is 2 inches, then to say they cannot be put together can mean either that the application of physical strength or of a machine cannot force them to fit (a clear empirical statement), or that they cannot fit so long as the one remains 3 inches and the other 2 inches. The difference in the grammars of "They can't be fitted" in these two cases is like that between "This piece of chalk is longer than that" and "A 3-inch piece of chalk is longer than a 2-inch piece". It is a rule about the use of "fitting" that it makes no sense to say 3 inches fits 2 inches. The difficulty is in using the word "can" in different ways, as "physically possible" and as "making no sense to say. . . ." The logical impossibility of fitting the two pieces seems of the same order as the physical impossibility, only more impossible! If one fixes the use of "apple" so that it excludes the use of "not" before it, then the impossibility of fitting the two is not like the impossibility of a physical fitting.

Lecture XI

To turn now to the relation between mathematical equations and tautologies. If mathematical equations are not tautologies, what is the relation between the two? There are two reasons for saying that $2+2=4$ is a tautology, (1) that it is not an experiential proposition, (2) that there is a tautology with which this equation is often mistaken: "If there are 2 things here and 2 things there, there are 4 things altogether". In Russell's notation this was expressed by using the identity sign.

Russell's notation gives rise to puzzlement because it makes identity appear to be a relation between two things. His symbolism for "There are 2 and only 2 things satisfying the function f" is: $(\exists x,y): fx . fy . x \neq y . (z) . fz . \supset . z=x \vee z=y$. We must distinguish the use of the sign of equality here from its use in arithmetic, where we can look at the sign in $a+b=n$ as part of a rule of substitution to the effect that instead of n we can write $a+b$. What is bad about Russell's notation is that it leads one to think there is such a proposition as $x=y$, or $x=x$. One can introduce a notation in which the identity sign as Russell used it can be abolished. Instead of writing "$(\exists x,y) fx . fy$", to which we have the right to add "$x=y$", we can make it a rule not to write signs of equality, but instead write one variable if one wants to talk of exactly one thing, two if one talks of two things. My notation for "There is only one thing satisfying 'f' " is $(\exists x)fx\,.\sim(\exists x,y)fx\,.fy$*. As an abbreviation of this I write $(E1x)fx$. With the elimination of the identity sign, such expressions as "$x=y$", "$x=x$", and "$(\exists x)x=x$" will not appear.

The symbolism used by Russell in which the identity sign occurs is puzzling because "$x=y$" and "$x=x$" seem to bring two objects, or an object and itself, into relation. "$(\exists x,y)fx.fy$ is a notation seemingly about things, but in saying x and y are identical we do not say anything about x and y; we want to say they are one. One might say that it simply means that the sign "x" means the same as the sign "y", but why is it that we should suddenly talk of the signs? Russell's notation is about the things referred to. If "$x=y$" can occur, so can "$(\exists x,y)\,x=y$". What does this mean? that there are two things that are the same? In my notation this is not a proposition at all, nor is $(\exists x)x=x$. Why, if there is one thing, should this be expressed by saying something about a thing? What tempts us to suppose that it is a fundamental truth that a thing is identical with itself (that this chair is identical with itself)?

I have not really done justice to the law of identity.

I shall now discuss the idea that "$1+1=2$" is an abbreviation of such statements as "If I have one apple in one hand, and another in the other, then I have two apples in both hands". In my notation this is: $\underline{(E1x)fx\,.\,(E1x)gx\,.\sim(\exists x)fx\,.\,gx\,.\supset.\,(E2x)fx \vee gx}$. (Recall that $(E2x)fx$ is short for $(\exists x,y)fx\,.fy\!:\,\sim(\exists x,y,z)fx\,.fy\,.fz\,.$) Now is it true that "$1+1=2$" is an abbreviation of the underlined? One thing to be noted is that what it abbreviates, if it does this, is much shorter than the corresponding expression in Russell's notation.

Suppose that instead of writing $(E15x)fx\,.\,(E27x)\,gx\,.\sim(\exists x)fx\,.\,gx\,.\supset.\,(E42x)fx \vee gx$, I wrote 56 in place of 42. Is this correct or incorrect? Suppose that among my rules of addition, some are definitions, e.g., $1+1=2$, $1+1+1=3$, and that others are deduced from them, e.g., $2+3=5$. If I wrote $2+3=6$, one *might* say this was not wrong, that it was merely a rule about the signs "$2+3$" and "6", so the effect is that for "$2+3$" I can put "6". But if it is called wrong you are already assuming a particular calculus; in another calculus it might not be wrong. The claim that $15+27=56$ is a contradiction may or may not be correct. How are we to find out? Suppose I try to find out

* See *Tractatus* 5.53–5.534 (Editor)

by putting it in the unabbreviated notation and determine by a calculation whether it is a tautology or a contradiction. To use a simple example:

$$(E^{||}x)fx.(E^{|||}x)gx.\sim(\exists x)fx.gx.\supset.(E^{|||||}x)fx \vee gx.$$

Whether this is a tautology or not I decide by *adding*. Now does it correspond to $2+3=5$? This implication says nothing (as it is either a tautology or a contradiction). But it would correspond to $2+3=5$ if to the unabbreviated notation were added "=Taut". What is queer about the functional notation $(E15x)fx.(E27x)gx.\sim(\exists x)fx.gx.\supset.(E42x)fx \vee gx$ is that we never use it when we are asked to reckon how many apples we have. One has to do an addition before one knows what to write after the quantifier in the consequent.

This leads directly to examination of Russell's and Frege's theory of cardinal numbers, of which the fundamental notion is correlation. Russell and Frege first introduced the idea of *being equal in number*. This was done via the notion of similarity, or 1–1 correlation. We shall take the commonsense point of view and call correlation anything like drawing lines, tying strings, holding hands. Two classes were said to be equal in number if they were correlated 1 – 1: ○ ○ ○ / × × × 3 was defined as the class of all triads correlated to ○ ○ ○. Any triad could be taken as the prototype, just as the Greenwich foot is taken as the prototype of all lengths of one foot.

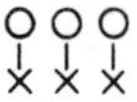

It is to be noted that Russell said that 2 is the class of all classes that *are* 1–1 correlated to [○ ○], not that it is the class that *can be* 1–1 correlated to the prototype. The latter is the amendment which everyone wishes to make. Suppose I removed cups from saucers, so that they were no longer correlated. Do they still have the same number? We would ordinarily say Yes. But how do we know this? That they do is now an *hypothesis*.

The question whether they have the same number if they *are* or if they *can be* correlated is bound up with whether the class is given in extension or not. Suppose I have two lists of letters

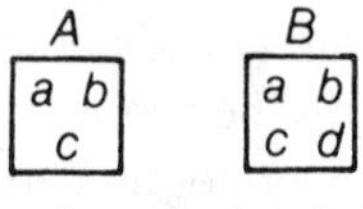

A	B
a b c	a b c d

The following two statements are not the same: (1) *abc* can be 1–1 correlated with *abcd*, (2) The letters on list A can be 1–1 correlated with those on B. For (2) can be decided by experiment, such as setting cups and saucers together, or drawing lines. This comes to a measurement of the number. There are other ways than 1–1 correlation for measuring the number, e.g., seeing that two geometrical shapes have the same number of intersections:

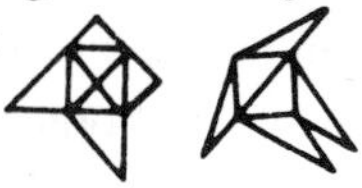

or seeing patterns of dots and crosses which are visually grasped as having the same number though no correlation is made: xx oo oo . (Beyond a certain small number this method is not available. If I saw 30 dots here and 30 dots there I might not be able to give their number, but I should be able to say whether one dot had vanished. This would not be the case if there were two pairs of 1000 dots.) Each method is a different way of determining whether two classes have the same number.*

Consider the difference between the criteria "can be 1–1 correlated" and "are 1–1 correlated". If the criterion is the possibility of two classes being correlated, we need to specify what role this possibility will play in determining whether they have the same number. Are we to say two classes have the same number when they have not been 1–1 correlated? or that two things have the same length when they are no longer superimposed? Sometimes yes, sometimes no. If we say they have the same number when they *can be* 1–1 correlated, we must fix the criteria for none having vanished. When presented with thousands of dots we do not know when some have vanished. But if we say two classes have the same number when they *are* 1–1 correlated, we do not have the question as to what happens when they are not correlated, nor do we need to take account of cases where we are not able to correlate. We can say that classes are equal in number if they *can be* correlated provided we give instructions for telling how we find whether they can be.

The criterion for sameness of number, namely, that the classes concerned *are* 1–1 correlated, is, however, peculiar. For no correlation seems to be made. Russell had a way of getting round this difficulty. No correlation need actually be made, since two things are always correlated with two others by identity. For there are two functions, the one satisfied only by a, b and the other only by c, d, namely, $x=a . \vee . x=b$ and $y=c . \vee . y=d$. By substituting a for x and c for y we have $a=a.\vee.a=b$ and $c=c.\vee.c=d$. We can then construct a

* See *Philosophische Grammatik*, p. 354. (Editor)

function satisfied only by the pairs *ac* and *bd,* that is, a function correlating one term of one group with one term of the other, namely, $x = a . y = c . \vee . x = b . y = d$, or the function, $x = a . \vee . y = d : x = b . \vee . y = c$. These correlate *a* with *c* and *b* with *d* by mere identity when there is no correlation by strings or other material correlation. But if "=" makes no sense, then it is no correlation. Why does this function seem to correlate them? Because of the identity sign.*

Lecture XII

To return to Russell's definition of a number as a *class* of all similar classes. A class can be represented in either of two different ways, (1) by a list, (2) by a common property. The class of men in this room, for example, might be represented by a list of their names. But Russell did not think of this class as represented by such a list, but by a property, *man in this room.* When Russell talks of a class he really means a property. He wanted to talk of a class in two ways. An existent class he wanted to talk about as a list, but he also wanted to be able to replace the list by a function. Frege had said a number is a property of a class. But he, and Russell, also said it was a property of a property. If there are five blue-eyed men in this room, 5 is a property of the property of being a blue-eyed man in this room. This account is unsatisfactory, however, since Frege also wanted to be able to say that Gans and Paul, for example, are two. And if 2 is the property of a property, Gans and Paul would be two only if they had a property in common, and one which nothing else had. There seems to be no reason why there should be such a property. Frege, however, thought he had found one, namely, $x = \text{Gans} . \vee . x = \text{Paul}$ (the property of being Gans or Paul). Frege and Russell thought they could manage classes intensionally because they thought they could convert a list into a property—a function. But if such a property, expressed by the identity sign, is objectionable, then what is meant by "Gans and Paul are two"?

Why were Frege and Russell so keen on defining number? In order to define it we shall of course have to define it in terms of other things we have not defined. Philosophers do not try to define everything, but certain things they have tried many times to define. What is common to those things for which they crave a definition? This craving arises from a question which bothers one and yet seems unanswerable in a

* See *Philosophische Grammatik,* p. 356. (Editor)

straightforward way. "What is a chair?", by comparison with "What is 3?", seems simple. For if one is asked what a chair is one can point to something or give some sort of description; but if asked what the number 3 is, one is at a loss. If one points to "3" and says that is the number 3, the reply will be that "3" is but a mark, of which the number is the meaning. This question, "What is 3?", arises from a jumble of misunderstandings, one of which is due to our having the word "meaning" in our language. "Meaning" is thought to stand for (1) something to which one can point, or (2) something in the mind. Suppose I ask whether the word "7" is meaningless in the sentence "There are 7 men in the room". Although it does not stand for something which can be pointed to,* everyone would reply that it is not meaningless, it is not superfluous. It has a function in the sentence. It is not the same as clearing-the-throat sounds. Although "function of a word" is not a definition of "meaning of a word", it is always useful to replace "meaning" by "function".

One great difficulty about numerals is due to the fact that they occur in utterly different contexts, in sentences of ordinary life, such as "There are seven men in the room", and in mathematical contexts such as "$2+3=5$" and "7 is a prime number". When people are asked, What is the number 3?, they first feel they are being asked to look about for something. Formalists, on finding nothing but the mark, said the number 3 *is* the mark. Others, on finding nothing when they looked about, said there must be something else than the mark. To those who maintain that the number 3 is nothing but the sign "3", the rejoinder is that one can have two shapes "3" and but one number. In fact we can say things about the sign that we cannot say about the number. Consider *the number 3* and *the sign "3"*. A person who says the number 3 is nothing but the sign "3" seems to say the two italicized expressions may replace each other. But they are not used in the same way. We can say that the sign "3" is red or written crookedly, but not that the number 3 is. Suppose I gave a name to the sign "3", say "dash". In place of "Write down a 3" the formalist would say "Write down 'dash' ". Now in answer to "What is dash?" I would point to the sign "3", but in answer to "What is 3?" I would not know what to point to. Here we see how the craving for a definition of

*Is it always the function of money that I get some material object for it? I might get permission to sit in a theatre, or a ride in a taxi, or extra speed. In each case I get something for 3 shillings, but its function is not always to buy an object.

number arises. On seeing that defining the number 3 as the sign won't do, we tend to say that since it is not the sign it is something else. The attempt to define the number 3 is like the attempt to define time. When we see that time cannot be defined as the movement of celestial bodies, we seek for another definition. Similarly for the king of chess. Since it cannot be defined as a piece of wood, we then ask, "What is it?" The craving for a definition of number is also prompted by the fact that in saying mathematics treats of *numbers,* and therefore of *the number 3,* as contrasted with using numerals in such contexts as "3 apples", the italicized expressions are *substantives**.

Suppose I gave "1 + 1 + 1 = 3" as the definition of "3", and am asked "What is 1 and what is + ?" The answer is that I can give their *use,* their grammar. Some people think that they say something by adding that "1" is not definable. But everything is definable, though not every word is defined in every game. It is sensible to ask whether it is defined in the game, and whether the definition is useful. I am not saying that a mark is the number 3 or that something else is, but that the explanation of the meaning of the word "3" is given by the use of the word "1" in the grammatical rule "1 + 1 + 1 = 3".

The question whether "3" has a meaning or is a meaningless mark arises because "3" has different uses in the sentences "There are 3 men here" and "2 + 1 = 3". Does "3" have a meaning in the first sentence but not in the second? We must understand the relation between a mathematical proposition about 3 and an ordinary one in which 3 occurs. The arithmetic sentence in which "3" occurs is a rule about the use of the word "3". The relation of this sentence to a sentence such as "There are 3 men here" is that between a rule of grammar about the word "3" and a sentence in which the word "3" is used. The application of a mathematical sentence occurring in our language is not to show us what is true or false but what is sense and what is nonsense. This holds for all mathematics—arithmetic, geometry, etc. For example, there are mathematical propositions about ellipses which show that "I cut the elliptical cake in 8 equal parts" does not make sense. And there are mathematical propositions about circles which show that it does make sense to say "I cut the circular cake in 8 equal parts". The terms "sense" and "nonsense", rather than the terms "true" and "false", bring out the relation of mathematical propositions to nonmathematical propositions.

*See *The Blue Book,* p. 1: ". . a substantive makes us look for a thing that corresponds to it". (Editor)

The question whether in "$2+1=3$" "3" has a meaning can be dealt with by examining a similar question about "$\sim$" in "$\sim\sim p=p$". I have said that "$\sim\sim p=p$" gives a rule for the use of the word "not"; it gives the meaning by stating how it functions. It is not about the mark "not" in the sense in which the sentence " 'not' is white" is about the mark. You might claim that it is about the mark in the sense in which "Mr. S is at liberty to use this mark in certain ways" is about the mark. But a rule does not state that one is allowed to do certain things. If it did, one might ask who allows us, when are we allowed, etc. It is misleading to say that a rule is a statement, a statement about a mark, for then there is a temptation to say it states that we in our society use a sign in such-and-such a way. If you say it is a statement about a mark, be careful. It is a rule about how the mark is to be used. Rules play a different role than statements. And we do not call a rule about "not" a command, although it is so used as to relate to commands, as well as to questions and statements.

Lecture XIII

I have been asked to give an analysis of the notion of a *rule*. Utterances might be divided into statements, commands, questions, exclamations, and perhaps instructions. Would rules fall under the latter classification, or is *rule* an ultimate classification? I would say that although in certain circumstances the above classification of utterances is useful (and such a cross-classification as *instruction* and *recipe* useless), it does not follow that classification is possible in other circumstances. Consider the blueprint which is produced in the drawing office and handed to the man at the lathe. Is it an order? It depends on its use. It could function as such, or might function as a statement, or as a suggestion. If he wanted to know how something was constructed, he might be given the blueprint to study. I should say that *rule* does not fit into any of the above classifications of utterances, that it is not in the same 'style'. The classification of utterances might be compared to the elementary classification of things in a child's reader into animals, human beings, foods, articles of furniture. If I were asked to classify a spintheriscope under one of these headings I should say "Roughly, it is an article of furniture", but it really does not fit any of these classifications at all; it is in a different style. Similarly, if required to classify $2+2=4$ within the rather primitive classification of utterances above I should say it is a statement, a statement about numbers.

What is peculiar about the classification into statements, questions, etc. is that each such heading is connected with a tone of voice. But a rule, as characterizing such things as $2+2=4$ as contrasted with rules for common games, has no particular tone of voice associated with it. We might say that rules *could* fall under all the divisions of this classification. Some might be commands, some statements, others instructions. But we would have to go on to specify under what conditions these function as rules. And there might be cases where rules do not fall within any of the divisions of the classification, any more than spintheriscopes belong to the classification *dress, food, furniture*. To call "$2+2=4$" a statement about the use of signs tends to lead to confusion. What I would call a statement about the use of signs would be of the form had by "All Englishmen use these signs in this way", or "All Chinese . . .", or "Hindus use this sign for the sun". It is confusing to regard rules as statements because this draws our attention to a different kind of question: Are they true or false? I emphasize the word "rule" *when I wish to oppose rules to something else,* e.g., when I wish to emphasize the difference between "$2+2=4$" and "If *A* gives me 2 apples and *B* gives me 2, then I have 4 apples in all". It does not follow that I can give you an explanation of what is common to all the things I do call rules. If inflection of voice is the basis for the classification of utterances, then I have made it impossible to include rules in it since they have no special intonation. Distinguishing utterances by tone of voice is a simple means in a language in which sentences are all written in exactly the same form but are uttered with different intonations. But in a language where in addition something might be a statement or a command according to the game it occurs in, the distinction would not be clear, as there would be two ways of distinguishing them, by tone of voice or by game. And there would be many more distinctions. If I made a classification which included rules it would have to be much more complicated.

Consider a game in which a table correlates letters with arrows indicating directions for walking,

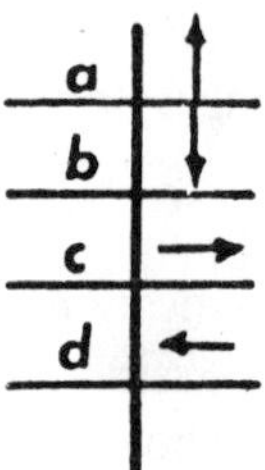

abbdccc would probably be called a rule, although the command to walk in such ways would not be. Were this chart used to make a frieze with a repeated design, one would be inclined to call it a rule. I should say a rule is something applied in many cases. If you said you have a rule when you have a certain kind of generality, I should say, Yes, so long as you realize that you have not said very much. It might be useful to say this in the case of a particular problem, e.g., when someone remarked that a rule that had been mentioned should be more general. But if I am a professor of logic and say a rule is something general or that for a rule generality is required, I am just making an ass of myself. For do you know any better how to use a rule from this explanation? It is quite useless; it tells you nothing.

The primitive classification into statements, questions, commands, etc. is useful because of certain obvious facts about human utterances, just as the classification exemplified in a family tree is useful because of the comparative simplicity of blood relationships. But if people were born in all sorts of different ways the genealogical classification would cease to be useful. Its usefulness depends on certain facts of natural history, and so does the classification of utterances.

A rule is best described as being like a garden path in which you are trained to walk, and which is convenient. You are taught arithmetic by a process of training, and this becomes one of the paths in which you walk. You are not compelled to do so, but you just do it.

Lecture XIV

Is $2+2=4$ a proposition *about* 2 and *about* 4? Compare this proposition with "There are no other men in this room than Jack and John". What is this about? Most people would say it is about Jack and John and the room. And of the proposition "There are no fairhaired people other than the two people Jack and John", they might say it is about fairhairedness, two, and negation. Compare this with $\sim\sim p=p$. Most people would say the latter is about negation, but if the sense in which the propositions about Jack and John are the paradigm for *being about*, then we would not say $\sim\sim p=p$ is about negation. Similarly, if "There are two men here" is taken to be about 2, then it is misleading to say $2+2=4$ is about 2; for it is "about" it in a different sense. We have said that $\sim\sim p=p$ is a rule about negation, and have also called $2+2=4$ a rule. The statement that I had two apples, that Johnson gave me two more, and that I ate the four apples is in accordance with this rule. It makes sense; whereas in light of the rule $4-5 \neq 1$, "I had four apples of which I gave away five and had one left" does not make sense.

If I told a mathematician that $2+2=4$ was a rule for the use of signs, he would feel uncomfortable. It has been said that $2+2=4$ is not a rule but a position in a game. We can have a game transforming $2+2=4$ into $2=4-2$ and $0=4-(2+2)$, these transformations being comparable to moving pieces on a chess board. Now why shouldn't we invent a game in which instead of using a chess board and pieces we have equations which we transform according to certain rules. The rules of this game might be the associative, distributive, and commutative laws of algebra. Given an equation, a sample problem would be to transform it in as few steps as possible into another form.

It has been said that $25 \times 24 = 600$ tells us a truth, given the multiplication rules, and therefore that we cannot say it is part of a game. It is not like chess, which gives no truths. But why should it not be a game? We could imagine a tribe which never made use of multiplication but played at multiplying huge figures, just as the Chinese never used gunpowder except for fireworks. Don't they really multiply? It might be said that what they do is not a real multiplication because they do not multiply in the right spirit to get a truth. But if $25 \times 24 = 600$ is a truth, would it be any less a truth because they were playing a game?

It has been said that $2+2=4$ is a rule for handling signs. But to say something is a rule of grammar is not to say it is always so used. The equation does stand to *propositions* using it in the relation of a rule to its applications. If you look at the use of what appears to be a statement you may find it is not a statement. Two statements, one of which appears to be about a state of mind and the other about a physical object, might be shown by their use to be entirely the same, e.g., the psychological statement "I believe the earth moves" and the astronomical one, "The earth moves". We could imagine a language in which every statement was preceded by "I believe"—as the German Swiss preface statements of fact by "Ich glaube". To discover of what sort a statement is we must examine how it is taught and learned, and how it is used in ordinary life.

Suppose we called "$2+2=4$" the expression of a convention. This is misleading, though the equation might originally have been the result of one. The situation with respect to it is comparable to the situation supposed in the Social Contract theory. We know that there was no actual contract, but it is as if such a contract had been made. Similarly for $2+2=4$: it is as if a convention had been made. And we can imagine a tribe acting according to the table of letters and arrows mentioned earlier without ever having been taught it. Like the table, $2+2=4$ is an instrument. The way in which it is taught us deprives it of all character as an utterance; it becomes impersonal.

To return now to Russell's definition of number. It has been said that number is the property of a class the members of which can be 1-1 correlated with a prototype, and also that it is the property a class has of *being* 1-1 correlated. The latter is not useless as a definition, as it gives one a way of finding out whether any set of objects has the same number as the paradigm. We might correlate objects by strings connecting them and the paradigm. One knows then what "I have 3 apples" means. Now Russell did not say we have three apples if they are correlatable to the paradigm, but that we have three if they are correlated. For he had a notion according to which similar classes are always correlated.

There is a tendency to say that when xxx can be correlated to the paradigm they *are* correlated, the idea being that the possibility of being correlated is like a thin thread joining the groups, and actual correlation like a thick one. A possibility is a shadow of reality. It is like the geometrical straight line which is so ethereal that it cannot be sensed. Russell's correlation is of this kind. It is a logical or possible correlation. And when it is said that a 1-1 correlation can be made though no physical correlation in fact exists, Russell says that nevertheless the two groups are correlated, as though geometrical lines join them. But a geometrical line is just the possibility of an actual line being drawn.

Two different expressions of Euclid's axiom, "A straight line, and only one, can be drawn between two points", and "One and only one straight line joins any two points", present the same question as do the two accounts of number. Let us examine the objection that any two points are not in fact joined by a line, and the reply that they are joined by a geometrical line, one which in contrast to a drawn line has no breadth. This sounds as though there is something there which could be made thick and gross. We must look at the relation between geometry and reality. If we say a geometrical line is drawn between two points, this means that it makes sense to say a physical line joins them, and that one and only one can be drawn means that it does not make sense to say two lines join them. The rule about what makes sense is equivalent to Euclid's axiom. The relation between the geometrical axiom and the coarse drawn line is the relation between a rule and its application.

To return now to Russell's ethereal correlation. To say two classes are correlated means that it makes sense to say they are. The classes *abc* and *def* are according to Russell logically correlated. (The word "logical" is like the word "geometrical".) The correlation of a and b is expressed in the formula $x=a.y=b$, and the correlation of *abc* and *def* by $x=a.y=d.\vee.x=b.y=e.\vee.x=c.y=f$. But how does one know they are correlated? One cannot know this, and thus whether they have the same number, *unless* one makes the correlation, i.e., by writing it down. Without doing this, to say the classes have the same number is like saying the ghost Finella has found them to be correlated.

Correlation by this relation serves nothing. We are cheated by the sign of equality. What does this sign in Russell's correlation formula mean? Can I try to see which things are identical with this chair? A form has been written down which sounds like a proposition. If a is put for x and b for y in $x=a.y=b$, the result is $a=a.b=b$. Now what does it mean to say $a=a.b=b$ is *true?* We've no use for this. By making such substitutions we have not shown anything about the number of a class. It is cheating, *except* that writing these equalities does make a correlation. What we *have* got is the result of a calculation.

Furthermore, the sign of equality when used in Russell's sense can be eliminated, in which case these equalities cannot be written down.

Note the difference between numerical equality between a class of nuts and a class of chairs, and between the sum of 2 and 2 and the number of roots of an equation of 4th degree, that is, between numerical equality outside mathematics and within mathematics. To one there corresponds a *measurement* of number and to the other a *calculation* of number. In the case of Russell's correlations we have a calculation and not a measurement.

Is there an experiment determining whether the two classes

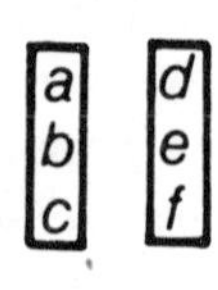

have the same number? There might or might not be an experiment in the case of classes which could not be surveyed. But if asked whether *abc* and *def* could have different numbers, the answer is No, since these can be surveyed. Would you call it an experiment to correlate *abcd....w* and $\alpha\beta\gamma\delta...\omega$ so as to *see* whether they have the same number? Would you say that you determine by experiment whether the number of numbers between 4 and 16 is the same as the number of those between 25 and 38? No, this is determined by a calculation, using dashes or something similar.

It is a pernicious prejudice to think that using dashes is an experiment and subtraction a calculation. This is comparable to supposing a Euclidean proof by using drawings is inexact whereas by using words it is not.

Lecture XV

There is a great difference between correlation in Russell's sense and in its ordinary sense: (1) the sense in which classes are correlated by means of *identity*, (2) the sense in which cups and saucers are correlated by placing one on the other. For in the latter case, to say they are not 1-1 correlated in this way does not mean they cannot be correlated in another way. But could one say the same thing about correlation in Russell's sense? Here correlation is such that if it does not hold, no other correlation could hold. This is the peculiar property of Russell's correlation relation as compared with ordinary correlation relations. I am not here calling attention to a phenomenon of nature; rather, to a matter of grammar. If correlation by identity does not hold, it does not make sense to say any other holds.

Frege's idea of *correspondence* does not necessarily imply any determinate correlation such as "sitting on" which might hold between three men and three chairs. One tends to say there is a correspondence between men and chairs whatever the physical facts may be—whether or not the men sit on the chairs. If there is a possibility of correlating them, there is a kind of attenuated correlation. This notion of correspondence, or correlation, is not taken from the physical world but from mathematics. The difference between physical and mathematical correlation is parallel to that between an experiment and a calculation.

To explain the difference between an experiment and a calculation let us consider the use of an abacus as an illustration of a calculation.

Keep in mind that calculation by an abacus is not less exact than one with digits. Let us work with the abacus and numerals, marking off one bead for 1, another for 2, and so on. Suppose that by moving the beads we make the calculation $2+3=5$. Now is this in fact a calculation or an experiment? This depends on the way we use it. Note that we *need* not get 5 beads, any more than that the specific weight of iron need be 7.5. It is quite possible for a proposition of experience to become a rule of grammar. Suppose experiment showed that something having all the other properties of iron had the specific weight 7.8. What would our attitude be to such a result? We might say it was a mistake. If it happened very often we might assign a different value to the specific weight of iron. Or we might hold that whatever the experimental results are, nothing is iron if it does not have the specific weight 7.5. In this case it becomes a rule of our language, whereas the proposition that the specific weight of iron is 7.5 was once an empirical proposition, confirmed at a given time and place. Similarly, if whenever we counted 2 and 3 and the result of addition was 4, we might say our rule must change. Or we might say that one of the beads had vanished, i.e., we might never alter the calculation $2+3=5$, though it might be very inconvenient not to. When we say $2+3$ *must* be 5, this shows that we have determined what is to count as correct; the *must* is a sign of a calculation. The difference between a calculation and an experiment is shown by our saying that a result of counting other than 5 is incorrect. When on counting two rows of apples we do not get the result calculated by adding their numbers, we can either say our addition rule must change or that the counting is incorrect. We would most likely say the latter. Or we might say one apple had vanished if the count was less than the calculated result. What is the criterion of an apple's vanishing? One criterion is seeing it vanish. But if we had two boxes of 25 and 16 apples, respectively, and after careful counting found only 40 apples even though we did not see one vanish, we might nevertheless say that one must have vanished. In this case we are taking $25+16=41$ as the criterion for one having vanished.

If we report that in counting with normal chalk $2+3$ always equals 5 but with Dover chalk 8, it is clear that we are talking of an experiment and are reporting a physical fact. But we could have started with either as a standard for judging experiments. Or we might accept both results and have different arithmetics. The facts do not compel us to accept one of them, but suggest the one we adopt. The connection between an empirical fact and whatever we lay down as a rule is that the proposition that conforms to fact is taken as the rule, other things being equal.

The fact that we can be mistaken in an arithmetic calculation is supposed to bear on the question whether arithmetic is an experiential science. But one is mistaken in a calculation for different reasons than in experimental determinations. "Mistake" is used in different ways in science and in arithmetic.

Lecture XVI

The requirement that if two classes are to have the same number an *actual* correlation must be made between their members is troubling. As a way out, let us construct a parallel to Russell's account, using as classes two sets of points in a plane. Let us say they are equal in number if there exist *geometrical* straight lines connecting the points 1-1. This seems to solve the problem, since the question of their number is now independent of whether lines are actually drawn. Geometrical lines, Frege claimed, always exist. This, however, is only an apparent way out of our trouble. For what is the criterion for there existing geometrical lines correlating the points? One might reply, "If I can draw connecting lines, then geometrical straight lines connecting them exist". But if one can show a way of connecting them by material lines, why say that geometrical straight lines exist? By saying that geometrical straight lines connect them I have only altered my expression of the criterion for their being 1-1 correlated. We have given no criterion for deciding whether the two classes are equal in number, for we can translate "A geometrical line correlates two points" as "It *makes sense* to say a real line is drawn between them". If the statement "the number of A = the number of B" means "It makes sense to say a 1-1 correlation is made" the assertion of numerical equality is a proposition of grammar, and says nothing about reality. To say that $10 \times 10 = 2 \times 50$ if the units are geometrically correlated is to assert a proposition of grammar; it is not about the world.

The charm of saying that geometrical straight lines connect the points is that we seem to say that a correlation exists. But the statement about geometrical lines does not say anything about reality. It does not mean that there is a correlation, but rather that it *makes sense* to say a correlation exists. Russell's theory, like this parallel account of correlation by geometrical straight lines, also makes it appear that a correlation already exists, and that it exists before it is determined experimentally. It makes it seem that we have reduced the question of numerical equality of classes to the question "Are they correlated?", before this question is settled experimentally.

Let us again look at the use of the word "can" in "can correlate". Suppose it is said that two points can be correlated by drawing lines between them. Whether one can do this depends on certain conditions—that one is not killed, that the surface does not vanish, etc. Certain conditions interest us and others do not. (Suppose the noses of all Englishmen could be joined to those of all Germans. What if the Germans refused? This is a possibility that does not interest us.) Now which is the interesting condition? Assuming that one will live and that no one hinders one, what condition remains which will enable one to draw the lines? No physical conditions seem to be of interest; for we say that whatever the physical conditions are, it is possible to draw a line between two points. By "possible" we mean *logically possible*. Where is the phenomenon of possibility to be looked for? Only in the symbolism we use. The essence of logical possibility is what is laid down in language. What is laid down depends on facts, but is not made true or false by them. What justifies a symbolism is its usefulness.

To talk about logical possibility is to talk about a rule for our expressions. Suppose that on counting two sets of dots several times over we get different numbers, and then that we 1-1 correlate them by strings. To say that our supposition is impossible shows that it is a rule that we are not to say both that they have different numbers and are 1-1 correlated. We are maintaining, at all costs, that if the two procedures do not hang together there is a mistake. If our joining them by lines were an experiment, what failed by one method might succeed with another. In this case we might say it was very *unlikely* that one should be able to join them by chalk lines and not by strings. But if we say this is *impossible*, it makes no sense to talk about making an experiment. The possibility of 1-1 correlation has to do with a symbolism. When I count • • and

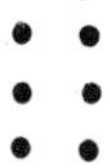

• •
• •

and say, "Yes, they can be correlated", have I come to a conclusion? No. I have said what is meant by connecting the dots of the two figures 1-1.

We have no need of a definition of number, and it was only thought that we do because "number" is a substantive which was regarded as denoting a thing with which mathematics deals. Russell's account of *having the same number* makes it appear to entail a correlation, a correlation of classes by an ethereal relation. This relation is really a chimera, and to say that classes are so correlated gets us no further than saying they have the same number. We cannot discover the logical correlation in any other way than by discovering whether they have the same number. If one asks what is the fundamental criterion for the possibility of 1-1 correlation, it is that they have the same number! Russell's definition of number is futile.

Easter Term

1935

Notes of lectures, supplemented by preparatory notes made by Wittgenstein

Lecture I

There is certainly something tempting in Russell's idea of number. But the idea of defining number at all springs from a misunderstanding. We do not *need* a definition of "number" any more than of "the king of chess". All a definition can do is to reduce the idea to a set of indefinables. And this was not the reason for which the definition was given; it would have been unimportant to do that. The reason was the insistent question, "*What is* a number?" We can get rid of the puzzlement of this question in a different way: by getting clear about the grammar of the word "number" and of the numerals. Don't ask for a definition; get clear about the grammar. By getting clear about the use of the word "number" we cease to ask the question "What is number?" Nor do we seek for something intangible which is, for example, the number 3, as contrasted with the digit "3". To observe that *the digit "3"* is not the same as *the number 3* only means that the italicized expressions have different uses. But if we crave a definition, then the definition Russell and Frege gave has a certain charm and it is understandable why it was such a success.

We have seen that the gravest difficulty with their definition comes out when one asks: Should we say we have 4 chairs if they *are* 1-1 correlated to a paradigm class, or if they *can be* 1-1 correlated? There is no reason why any group should be materially correlated with any other. But if not correlated materially, Russell and Frege wanted to say that in some ethereal way they are correlated. If the correlation is a drawn line, one feels that before the correlation there was the possibility—like a very thin line which one traces in heavily when one draws, or like a poem muttered quickly when one is asked whether one knows it from memory, which is then traced in heavy lines by reciting it. The possibility of correlation seems to be some sort of correlation. Often its being possible to do something is like doing something similar.

We must distinguish physical possibility and impossibility from the possibility and impossibility in which we are interested. The impossibility of correlating - - / - - and o o / o o because one becomes paralyzed while drawing lines between the members of the two groups is very different from the impossibility of correlating - - - and o o, where one member is left over when connecting lines are drawn. The latter has nothing to do with physical impossibility. What makes us call both of these impossibilities? They differ in that in one case what one person does not succeed in doing, another may. Here you might say, "My hand cramps. You try". In the other case you say, "Don't try. It's hopeless". In what sense is it hopeless? Suppose we have a group of *many* crosses which we begin correlating to a paradigm, neither of which we can survey. If one cross is left over we say, "Therefore, they cannot be correlated". What would we call trying to correlate them 1-1? Perhaps tracing the lines between paradigm and group again and again. Or someone might try, and claim to have succeeded in, correlating like this: x—o / x—o / x, x ⋝ o. We would object that he is not playing the game, and what we would mean is that it does not make sense to say that one succeeds or can try, or that the × 's and o's are connected or are not connected.

It is queer that one should be able to say "It is hopeless to try". This sounds like a prophecy, as does predicting that anyone who tries to get a 4 when he divides 1 by 3 will fail. But if it is a *prophecy,* it can be wrong. And it is a prophecy *unless* it has been *decided,* when one x is left over in the correlation, that it is impossible. The impossibility of correlating them 1-1, or of getting a 4 in the division, does seem to be similar to another sort of impossibility—to a physical impossibility or an impossibility of remembering. I want to destroy this seeming similarity. Compare the prophecy that it is hopeless for me to win a fight against a heavyweight boxer with what seems to be a prophecy of what will happen when I try to correlate x x x and o o. How do you know that I will fail? In the first case it is because the boxer is bigger than me. Note, however, that here I could give a description or have a picture painted of such an imaginary fight in which I did (or did not) win, as well as of a real fight. But *can we describe what we cannot do when we try to 1-1 correlate* $\begin{smallmatrix}\times\\\times\\\times\\\times\end{smallmatrix}$ *and* $\begin{smallmatrix}\circ\\\circ\\\circ\end{smallmatrix}$ *or get a 4 in the result of dividing 1 by 3? Anything that can be described can happen.* If we describe correlating $\begin{smallmatrix}\times\\\times\\\times\\\times\end{smallmatrix}$ $\begin{smallmatrix}\circ\\\circ\\\circ\end{smallmatrix}$ we *can* correlate them. There is nothing to prevent our joining them in any way we wish, but we must fix what is to be done. Instead of saying that $\begin{smallmatrix}\times\\\times\\\times\\\times\end{smallmatrix}$ can never be correlated 1-1 to $\begin{smallmatrix}\circ\\\circ\\\circ\end{smallmatrix}$ one should say "No correlation of these two groups will be called 1-1 correlation". It is a rule we give.

Trying to catch one's thumb is similar to trying to correlate $\begin{smallmatrix}\times\\\times\end{smallmatrix}$ and $\circ$. There is a conflict between the aim of a person who wants to catch his thumb and the fact that he would not be satisfied had he done it. One might deceive him by putting a replica in place of his thumb when his hand moved, but on discovering the deception he would say this is not what he was trying to catch. He is like the person who would not be satisfied had he correlated $\begin{smallmatrix}\times\\\times\end{smallmatrix}$ and $\circ$ by some game of skill.

Let us compare "It is impossible to get from this room to the next without going through the open door" with "It is impossible to catch one's own thumb". In the first case you can describe both beginning and end and the condition without which the next room cannot be reached. In the second case, you have not said what it is that is impossible, i.e., what you are not going to succeed in doing, for there is no describing catching one's thumb. When you tell someone he is not able to do a certain thing a muddle arises if he thinks that you have told him *what* he is incapable of doing. "1-1 correlating four crosses with three circles" does not describe what it is impossible for him to achieve.

To return to the identity relation which Russell said made the correlation between classes. That any two things A and C are correlated 1-1 shows up by replacing x by A and y by C in the function $x = A \cdot y = C$. "$A = A . C = C$" is supposed to assert the relation between A and C, and we ought to be able to find out whether the two groups have the same number by seeing whether this identity relation holds between their members. Suppose the people in two rooms are named by the letters of the Latin and Greek alphabets, respectively. It would seem that by calculating we could get to know that the number of people in the two rooms is the same: A corresponds to α, B to β, etc. Now on writing the formula $x = A . y = \alpha . \vee . x = B . y = \beta$, $\vee$ . . . . etc. we have only correlated the names. What we have really done is to find out that the Greek and Latin alphabets have the same number of letters! Is there any other result? We might say that through the names we have correlated the people; and if a Greek or Latin letter is pinned on each person we might be said to have done so. But if one letter were given to several people, if A were given to x and B were given to xxx, Russell's formula would not show that the two groups of people have the same number. We might say that by correlating the names one knows how many *entities* there are; but what good is this if one does not know what the entities are? If into the schema $x =\quad .y =\quad . \vee . x =\quad .y =\quad . \vee$ . . . . . I write names, I shall have correlated the same number of names with the x's as with the y's. If I *then* correlate things with the names I can say that the things have the same number as the two groups of names. But this is a purely material correlation.

Lecture II

"For every one here there is one there". These words say nothing about any actual relation between the two groups and seems to be independent of any actual correlation between them. It is conceived in terms of checking one item against another. Were Russell's definition useful it would give a hint as to a method of discovering whether two classes have the same number. We use the words "for this there is this" when we deal with *names* of things, and do not make a real relation between the things. The things denoted by "*ABC*" and "*DEF*" are said to have the same number if they fall into couples, but all the couples are held to exist whether the pairing is done or not. It is when we talk of the symbolism that there is a temptation to talk of a correspondence which is not an actual correlation. To see whether the two classes have the same number, put them in 1-1 correlation. An actual correlation does not give us an indirect way of finding out whether they have the same number, but tells us what we mean by classes having the same number.

Determining whether two classes have the same number may be either an experiment or a calculation. Let us compare cases where we can see immediately that the number is the same, or different, and cases when the number is too large for this. In the case of $\circ\,^{\times}_{\times}$, determination of whether circles and crosses are numerically equal is not an experiment. It is a calculation, like finding out whether 5′ is more than 4′9″. For a greater number, where a visual criterion would not serve, it would be an experiment. There would be no such thing as seeing the numerical difference between 1000 and 1001. We can regard $\begin{smallmatrix} \circ & \times \\ \circ & \times \\ & \times \end{smallmatrix}$ in two ways. (1) We can survey all of them here—look at the array as a whole, as one picture—and see that the two groups are numerically different, inasmuch as there is one isolated cross. (2) We can survey them after making a correlation to see whether any element remains over, say, by covering each row of circles and crosses successively, finally uncovering, as the last stage of an experiment, a row with a cross left over. Here we have an experiment, not the "result" of a demonstration. The experiment shows one cross remaining. Only if the two classes are looked at as a whole, and one cross is seen as isolated (with no possibility of a cross having disappeared), would it be a demonstration.

Lecture III

Russell's relation "I am I" and "he is he", which correlated *I* and *he* when one substitutes in $x = \mathrm{I} . y = \mathrm{he}$, is popularly expressed by "for this there is this". Saying that for this there is this sounds very like saying that they stand in a certain relation—such as "This is the wife of this man and this the wife of that man". "For this there is this" really says nothing about the terms, or a relation. It has the form of a proposition about things, and at the same time is not one.* Likewise to say that two classes have the same number of members if the members fall into pairs says nothing, but it sounds like a real statement such as "These men fall into pairs", meaning that they walk together. Having the form of a proposition about things and yet not being one gives rise to muddle.

It is important to remember that when we talk of classes having the same number of members we by no means always talk of the same phenomenon. Compare saying that two coal scuttles have the same number of pieces of coal with saying that the two rows $\begin{smallmatrix} \times & \circ \\ \times & \circ \\ \times & \circ \end{smallmatrix}$ have the

* #These words, together with gestures, are being used to put the members of the classes into couples. And this *is* a way of finding whether they have the same number.# (From Wittgenstein's preparatory notes. In future, supplements from his notes will be indicated by #. (Editor)

same number of elements. You might say that we can ascertain whether the two scuttles have the same number of pieces by counting them. But I might object that I want to know how many pieces there are *now,* and that emptying the scuttles piece by piece does not tell me that there will be the same number after counting as there were when I asked. How does one know that some will not vanish while being counted or that others will not break up? Similarly, suppose someone asks me whether two pieces of wood are the same length, and I put them together and say, Yes. Suppose he says that he wanted to know whether they are the same as when he asked me, and questions whether I know that nothing has happened to them while I measured them which would make them a different length now than then. The answer is that it means nothing to say they are the same if every method of finding out is rejected. He should be asked what his criterion is for their being the same length. There may be many criteria. To say they have the same length is to say something about the method of finding out. The same applies to numerical equality. A statement about the number of terms means different things according to the criterion for finding out what the number is.

Suppose I take as criterion for the equality of numbers what our explanation gives, namely 1-1 correlation. To say the numbers are the same is to say something with many different grammars. Note how different is 1-1 correlation of coals in the two scuttles and of the crosses and circles in $\begin{smallmatrix}\times & \circ \\ \times & \circ \\ \times & \circ\end{smallmatrix}$. In the latter case one has a visual criterion; one sees them correlated. Where there is a visual paradigm one has a method for determining whether two classes have the same number. A large paradigm cannot be surveyed, so that this method fails beyond a certain number. The phenomenon of 1-1 correlation is entirely different in different cases. With the coals in the two scuttles we do not have a visual criterion. Yet we call 1-1 correlation both visual correlation, in the case where one does not ask whether the two groups have the same number, and throwing coals out of the window in pairs, where no visual correlation exists. In the case of the coal scuttles, we use correlation to find out that they have the same number of coals. But in the case of a visual paradigm like $\begin{smallmatrix}\times & \circ \\ \times & \circ \\ \times & \circ\end{smallmatrix}$, which is a most special case, we can both see the pairings *and* the number. Here correlation is not a criterion for the two sides of the array having the same number. Sometimes we cannot see either the correlation or that they have the same number. In the case of the coals thrown out in pairs, we can do only one thing: use correlation. The idea of 1-1 correlation is an image through which we look at a certain fact, and in some cases is most natural, namely those in which there is a possibility of joining terms into couples. For some cases it is very suitable, and not at all for others. Recognition of the same pattern is another criterion for numerical equality. One can say that "There are four people in this room" means that the people can be put in the form of a square, and so on for other numbers, as follows

```
• •        •
 •        • •
• •      • • •
        • • • •
```

These schemata are natural but are very limited. We could not count up to 100 by them. The grammar of "Two classes have the same number" differs in different circumstances. And to say that two waves have the same number of nodes when there is no way of correlating them means nothing.

The pattern

```
L R
  o
x o
x o
```

may be regarded as a *demonstration* of $3>2$, one which is as good as any geometical demonstration. (To recognize that this is so I should have to give you a system of such demonstrations by which you would see that what I did was an operation laid down beforehand, like that for addition and subtraction.) However, if we look upon it as a pattern on which we *experiment,* then it is not a mathematical proposition to say there are more elements in R than in L. What one calls a *demonstration* does not have a proposition as a result, as does an experiment, but instead a rule of grammar.

When I draw a circumscribed pentagram

and find that the outer vertices are five, have I made an experiment? Is determining that the number of outer vertices of the figure is the same as the number of inner vertices an experiment? We must distinguish here between "This figure has 10 vertices" and "The pentagram has 10 vertices". Of course drawing the lines of this figure, to get the result that there are 10 vertices, is an experiment. Drawing the figure takes place in time. But the *demonstration* about *the pentagram* is not what I did at a particular time, in these five minutes of drawing; it is not a process at all. It is just the figure. The proposition about *the pentagram* is a rule of grammar about the word "pentagram".

It is said that if one demonstrates something about this pentagram one has demonstrated the same thing for all pentagrams. This is a confusion. What is wrong here is the word "this". A demonstration is not about *this* figure. Finding out something about this particular pentagram is an experiment; whereas a demonstration shows something about the use of the word "pentagram". It is nonsense that an experiment with this figure should be used to demonstrate something about this figure.

There seem to be three propositions which can be asserted about this figure:

(1) These sets of dots do not fall into pairs. This is an empirical proposition. (2) Three dots cannot fall into pairs with four dots. This is an arithmetical proposition. The phrase "falling into pairs" is used in a timeless way having nothing to do with a *method* of pairing. "4 dots do not fall into pairs with 3 dots" is timeless. (3) Anything that looks like this  cannot fall into pairs with anything that looks like this

Here we seem to have proved something about reality.

Lecture IV

Existential statements about classes have very different grammars. For example, "There is a 1-1 correlation between the members of *A* and *B*". #We have said that what we described as "numerical equality", "being 1-1 correlated", "having the number *n*" were widely different phenomena. It is an illusion to think that "members of the classes fall into pairs" is an *analysis* into simpler terms of what we call numerical equality. We can if we like put "being numerically equal" = "falling into pairs", but the use of the one expression just as of the other has got to be explained in the particular case.# To say that two classes have the same number and to say they can be 1-1 correlated are two verbal forms and nothing else. We do not know how either expression is to be used in a particular case. If there is a way of correlating the classes, then in saying that "having the same number" means being 1-1 correlated we refer to a process already used; and thus the definition sounds illuminating. But unless there is a way, we would do better to define "being 1-1 correlated" as *having the same number*. This might be the case if we were in a position to count but not to 1-1 correlate. Our definition gives a very convenient formulation, not for every case, but for those cases most likely to be thought of. Similarly, "*A* is double the length of *B*" defined as "*B* superimposed on *A* twice gives the length of *A*" is convenient as applied to rods. But it would be senseless as applied to wave lengths of light. For it has not been said what it would mean to put together two wave lengths and find that one is double the other.

Let us compare "The pentagram has 5 outer vertices" and "My hand has 5 fingers". These are enormously different, although they sound alike. Their grammars differ in a way that could be described in terms of ordinary English grammar: the first has *no tenses,* whereas one can say of a hand that it *has had* 5 fingers. The proposition which answers the question "How many?" is in the first case timeless, though there is a great temptation to say that a pentagram *always* has 5 outer vertices. Apart from a context you might think the question meant "How many does it have now?" as against "How many did it have yesterday?" The question asked about the pentagram's vertices is not about a specific figure drawn on the blackboard.

Suppose the human hand were taken as the sample for 5. Then the statement that the hand has 5 fingers would be like that about the pentagram, or like the statement that there are 12 inches in a foot. It would be of this timeless kind. When one asks in the mathematical sense how many vertices the pentagram has, one has resolved that one particular cardinal number be correlated with it. And it is not said *which* pentagram this number is correlated with. This is not the case with the fingers of a hand. We might, however, have different conventions for counting. We might say the pentagram has six vertices because we had decided to count one vertex twice over. And we might have different conventions for linear counting and for counting in a circle. This would not make havoc in arithmetic. We should simply have a different arithmetic. There is no abstract reason for correlating one particular number with the pentagram's vertices. In fact we could continue counting vertices as long as we wished, in which case it would not make sense to ask how many vertices there were.

In saying the number of vertices is five we might think what we are saying is in no way dependent on a convention, that we have said something about the *nature* of the pentagram. People have thought that although it is not in the nature of a hand to have five fingers, it is in the nature of the pentagram to have five vertices. We are also inclined to think that in a demonstration about we have demonstrated something in its essence. This is a dangerous trap. Consider

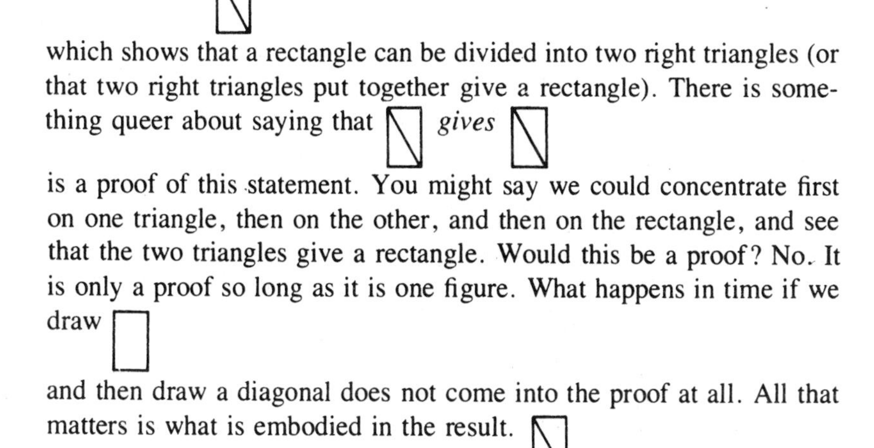

which shows that a rectangle can be divided into two right triangles (or that two right triangles put together give a rectangle). There is something queer about saying that *gives* is a proof of this statement. You might say we could concentrate first on one triangle, then on the other, and then on the rectangle, and see that the two triangles give a rectangle. Would this be a proof? No. It is only a proof so long as it is one figure. What happens in time if we draw and then draw a diagonal does not come into the proof at all. All that matters is what is embodied in the result.

If the diagonal suddenly becomes wavy, this makes no difference, as what happens in time within the diagram is irrelevant. "This gives that" is ambiguous. One diagram and another do not give a third in the way in which hydrogen and oxygen give water. The diagram showing that the two triangles give a rectangle, in being described by the phrase "This and this give that", is a standard for judging any further experience.

Let us look at the simple demonstration that $2+2=4$: ⊔⊔ Here also it appears that two diagrams—this and this together—give 4. But nothing is happening at all. There is just the figure. Suppose one said of |||| that it is in its nature to be divided into two 2's. The numerous ways of dividing these strokes mislead us; they make us say that it is in their nature to be capable of being divided. These strokes, ||||, the image of 4, presents us with a queer case: it seemingly has something in it from experience and something *a priori*. We seem to have demonstrated that the visual 4 is capable of being divided into couples, that ⊔⊔ demonstrates that |||| can be so divided. But which four strokes consist of two 2's? |||| or ⊔⊔? You might say that the figure consists of two 2's when it is divided into two 2's, but does not do so until it is divided. And then what you see is the whole drawing, like one whole experiment of which you see the beginning and end in one figure. However, it is not this; it is a picture of what an experiment would be if it gave this result. The picture of an experiment is not an experiment at all. But this picture could be used for two entirely different purposes, (1) to show what the experiment yields, i.e., to describe what actually happens. A film which shows  shows three stages of what an experiment yields when strokes are grouped and an arc added. (2) to serve as a standard for a normal experiment. For example, if a picture was drawn of a flower and we said, "This is not the picture of a real flower, but if a flower grows in this way we shall say it grows normally", then we are using the picture as a standard. The diagram  supplies us with a standard: if two things put together with two others give us four, then we shall say nothing has been added and nothing has vanished.

In some cases it is not clear what is demonstration and what is experiment. Consider Pythagoras' theorem. It probably was not a theorem at first but a proposition of experience. By measurement it was found that the sum of the squares on the sides was equal to the square on the hypotenuse.

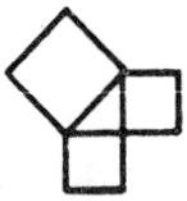

Here one seems to have a proposition playing two different roles, (a) found true in all cases, (b) demonstrated. And it is queer that what is found true in particular cases should be demonstrated. It is as if the table stood on more legs than is necessary.

What makes it difficult for us to understand as a demonstration of $2+2=4$ is in the idea of the visual 4. We might say we see one figure and the other figure and then both together, but that we do not see a "must". What do we say the proof is about? Is it about 4, or about three 4's.; about |||| or about  ? (When I have written AA on the board, have I written one letter or two? This depends on whether I mean by "A" this peculiar shape or a collection of chalk marks.)

Suppose I wish to lay boards 2 feet wide across a room 12 feet wide. Arithmetic tells me I shall need six boards to cover it. I get six such boards and say, "Therefore they will fill the room". But need they? If they do not, I shall say they must have expanded or contracted. This means that arithmetic has told me nothing. And though it appears that arithmetic has told me something if I find on laying six boards that they are each 2 feet wide, again it has not told me anything. For now I call 2 feet what goes 6 times into 12.

Suppose I am doubtful whether three boards 2 feet wide fill 6 feet. This picture seems to demonstrate it

and I might say it shows something about the nature of 6, namely, that it is 3×2. There is no experiment here, and nothing is shown about the nature of 6. We must distinguish between "This space can be covered by three boards 2 feet wide" and "A space of 6 feet can be covered by three boards 2 feet wide". The latter is not demonstrated by an experiment.

Lecture V

Suppose a space were fitted by three boards 2 feet wide. There is no difficulty in seeing that the mathematical proposition $2 \times 3 = 6$ says nothing about the boards or about their fitting the space. For if we filled the space with boards supposedly 2 feet wide and they did not fit, we would say either that the boards had changed or that we had measured wrongly. Mathematical propositions do not predict, not even that they will fit if they do not change. For the criterion of the boards having changed is that they were measured correctly but do not fit.

I have been trying to deal with visual geometry where proofs consist of looking at figures and seeing connections between them, what one might call visual demonstrations. Consider (a), , and (b) . Does (a) fit (b)? We place (b) under (a) and see that they do fit. It seems that we can show this by something which is like an experiment and yet not an experiment. It is something like a demonstration about reality. The question about (a) and (b) is of the same kind as the question whether fits

By joining the vertices of a pentagon we get a pentagram, and show how they fit. (I take figures here which one can see as ornaments and recognize. It won't do to take figures with 20—32 divisions.) Note

Does a hexagon fit an equilateral triangle? Is it an experiment to prolong the faces of a crystal so as to get a figure that surprises one, e.g., to extend some faces of an octahedron and get a tetrahedron? All are the same sort of question. If one is an experiment, the other is. There *is* one matter of experience, viz., that by drawing lines in such-and-such a way one gets a well-known figure, say the pentagram. But it is not a matter of experience that the pentagram fits the pentagon. There is a peculiarity about the word "fits" here; it is tenseless. What 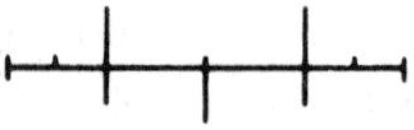 shows is not that the figures (a) and (b) fit, but *what it looks like* for these figures to fit. One might say that the complete figure

shows what one means by a pentagram fitting a pentagon.

I want to free you from the idea that this figure

shows something about the essence of a pentagon and a pentagram, that the demonstration is an experiment in another sphere, an experiment at a higher level on a more ethereal object.

Suppose we had never seen a pentagram inscribed in a pentagon, that it is a new experience. #We are tempted to say that the experience teaches us that the pentagram fits the pentagon. It seems that we are learning by experience a timeless truth.# At the same time we tend to think the experience of inscribing the one within the other is an experiment with sense data, and that it proves something about things in our visual field. This is wrong. #What is important is that although "visual image *p* (the pentagram) fits visual image *P* (the pentagon)" seems to be proved by experience, it is used as a proposition of geometry, i.e., of grammar.#

In what cases do we call a color a blend of two other colors? When one mixes blue and white paint the result is pale blue. But what if mixing them gave green? We would say that could not be, that a chemical reaction had taken place. Consider mixing light, which does not introduce the complications which paint does. Suppose mixing red and blue gave yellow instead of purple, and that it did this even on the color wheel. The color wheel can be used to experiment and also to demonstrate. In some cases it is not used to make an experiment but to show us when we are getting a real blend of colors, i.e., when they are mixed normally. If something different happens from what we expect we say something must be wrong. We *could* say, "Whatever the color wheel gives when red and blue are mixed, I'll call that the blend of the two colors", but we do not in fact say this. As when the colors of the rainbow are mixed and give grey instead of white, so if a color wheel gave an unexpected result we should say the colors mixed were not quite pure. When we use the color wheel to make an experiment, and say "Blue and white give pale blue", we can say it does this *now*, and "give" is temporal. Otherwise used, "This is a blend of blue and white" or "This . . . gives this . . ." is not a temporal statement, just as "This . . . fits this . . .", similarly used, is not temporal.

Lecture VI

It is easy to see that statements of geometry and arithmetic are used as statements of grammar, that they do not predict, for example, what will happen when two rods 2 feet long are put together and set against a third rod of 4 feet.

It does not follow at all that by putting the two-foot rods together one will get 4 feet. If one says they *must* give us 4 feet, that does not say anything about the actual result of measurements. One might (wrongly) argue that all that need be done to show that the two rods equal a rod of 4 feet is to divide the latter in this way:

and conclude that *a must be* 2 feet. This assumes too much, for how do you know that if the lower right-hand rod is 2 feet the left-hand length *a* must be 2 feet? The total length may well be 4 feet by construction but not by measurement. Similarly, a construction by compasses of a regular pentagon in a circle may show the sides to be equal, but measurement may not show this. If *a* is 2 feet *by definition*, then the whole matter is different. That is, if the *measurement* is to be confirmed by the calculation, then one can see that $2' + 2' = 4'$ is a grammatical proposition and not about actual lengths. There is the appearance, however, that a proposition of grammar is confirmed by experience, an experience, furthermore, of an internal relation.

Let us ask the following mathematical question: What sort of star do the diagonals of a pentagon form? The answer is given by drawing a pentagon and its diagonals, then isolating the inscribed figure. Here we have a problem and also its solution. An exactly analogous question is the following: How much is 23×18? The calculation

$$\begin{array}{r} 23 \\ \underline{18} \\ 184 \\ \underline{23} \\ 414 \end{array}$$

is as much a geometrical construction as

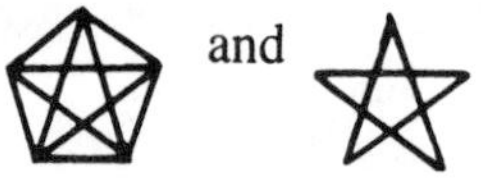

and

The multiplication table might be regarded as a means of constructing the figure. The following is also a geometrical construction, showing that $3 \times 3 = (4 \times 2) + 1$. If I call drawing the star which is formed by the diagonals of a pentagon an illustration of a geometrical construction, then the multiplication is also. But I should prefer to say that if the latter is a calculation then so is the former.

It has been said that constructions with ruler and compasses are always inaccurate since one can never draw a geometrical straight line, and hence the drawing

is not exact. This is not a proper objection, but if it were, the same would apply to the multiplication: it could be objected that the shapes of "4" were not exact, that we could never be sure we had written "the arithmetical 4".

If we say the construction shows us something about the essence of pentagrams and pentagons, then we must say the multiplication shows something about the essence of numbers. Would it be wise to say this? Suppose we taught a man multiplication, giving him rules for multiplying decimal numbers. Suppose that besides these rules we assumed a further, queer axiom: $23 \times 18 = 800$. Is it wrong to assume this or not? Is there a sense in which an arithmetic which had this is no arithmetic? Some might say it would be uninteresting, because from a contradiction together with the usual rules one could deduce anything. I say this is wrong. Assuming the usual rules of arithmetic, was I bound to get the result 414 from 23×18? The proposed axiom, $23 \times 18 = 800$, is uninteresting because there is no phenomenon for which an arithmetic including it is of any use. If any phenomenon arose, i.e., if 23, 18, 800 were constants relating to our space or to all natural phenomena, it might be extremely useful to have an arithmetic in which one multiplication among all others had two results. What about the objection that if we had this rule we should have to add others? If we taught a man a rule by giving him instances up to 500, expecting him to add thereafter according to the rule of "Add 2", and if at 1000 instead of adding 2 he did what we would call adding 3, what would we say if he claimed to have followed the rule? We must not suppose that with the rule we have given the infinite extension of its application. Every new step in a calculation *is a fresh step*. In answer to the objection above, the reply is that we need not give new rules. It is not in the nature of 23 and 18 to give 414 when multiplied, nor even in the nature of the rules. We do it that way, that is all. This does not mean that *any* result in arithmetic will be accepted.

Consider the following: "If you draw the diagonals of a pentagon you get a pentagram, and "if you do this and this . . . you get Napoleon."

What can't be predicted when the operations are described is the character of the visual impression. An experimental factor is involved. We ought to examine what is an experiment in these cases, and in such analogous cases as

containing  , and the puzzle picture of foliage in which the face of a man is hidden. What sort of proposition is it to say

contains 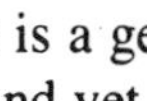 ? It may or may not be a proposition of experience. We are inclined to say that when the face of the man is seen when the foliage is looked at from a certain angle we have discovered something internally related to the foliage. One cannot draw the foliage without drawing the man's face, so that to say there is a face there is redundant. It is a geometrical proposition that in the foliage there is a man's face. And yet it is a new experience when one sees the face. #Our visual impression has changed. But must one not say that the new experience would have been *impossible* if the old one had not been what it was? We seem bound to say that the new experience was already preformed in the old one, or that we had found something new which was already in the essence of the first picture.#

What does one discover when one discovers the man's face or  in  , or that two right isosceles triangles put together give a rectangle? A new experience is involved, an experience of a *new aspect*. We say "Oh, that has never struck me; but I now see it must be so". We do not say this in the case of a genuine experiment.

Seeing a formula in a different aspect is sometimes regarded as a mathematical discovery. Sheffer discovered that Russell's two primitive ideas "or" and "not" could be defined in terms of one constant "neither . . . nor"*. $\sim q$ was defined as q/q and $p \vee q$ as $p/q \cdot / \cdot p/q$. Imagine that Russell and Whitehead had written *Principia* in such a way that "$\sim$" and "$\cdot$" were always distributed in the order $\sim \cdot \sim$, as in $\sim p \cdot \sim q$. And suppose that Sheffer discovered what they unwittingly did. One could say he was merely drawing attention to a certain aspect of the formula which they wrote in this fashion, just as the two triangles could be regarded as a new aspect of the rectangle.

Lecture VII

Let us look at the role which an *aspect* plays in mathematical demonstration, the aspect under which one sees figures, when something timeless seems to have struck us.

Consider the statement that a certain object consists of parts, e.g., that a pentagram consists of a pentagon and five triangles, and that a chair consists of back, legs, rungs, etc. The word "consists", which means "made up of parts", is used in two different ways. To say a chessboard consists of 32 white squares and 32 blacks gives information which a man might not have had. With this meaning given to "consists" he would be able to make a chessboard.

The parts a thing has depend on the different ways of dividing it. All sorts of things could be called dividing or putting together, one of which is visual division. A figure can be divided differently by different acts of attention. Suppose we visually divide IIII into two groups of two. One might say that this is *seeing* that $2+2=4$. But the equation can't be seen. There is no phenomenon of seeing that $2+2$ means the same as 4. But there is a phenomenon of seeing certain aspects. Seeing the dashes in pairs *suggests* a rule, $2+2=4$. A certain symbolism readily goes with a certain aspect which strikes us when we look at a thing.

Suppose a tribe saw a square as two intersecting parallels, and that another tribe always saw it as a double right angle: According to

*This is an alternative to the usual reading of Sheffer's stroke. (Editor)

the visual aspect they would very likely adopt a certain description, e.g., the use of "double right angle" instead of "square". But the description is not *necessarily* bound up with the aspect. And it is not the case that they *must* see it like this because they adopt the symbolism.

We can imagine a language which never used "4" but only "2 + 2". As long as we look at | | | | in "the 2 and 2 way" our picture consists of a division into 2 and 2. The actual visual division is a temporal process, and the figure will consist in division into two parts as long as the phenomenon of division lasts. But the equation $2+2=4$ is timeless. That 4 consists of 2 and 2 in the sense of "$2+2=4$" cannot be *seen*. There is no phenomenon of seeing that a proposition of grammar holds.

Visual division is a phenomenon like any other. If one does not recognize this one feels that by getting a new aspect one penetrates into the essence of the thing. When one's attention is drawn to the fact that a pentagram consists of a pentagon and five triangles one seems to see something which is there whether one's attention is drawn to it or not. However, that the pentagram consists of these parts lasts as long as we see it under this aspect. On the other hand, if we adopt the geometrical expression, "pentagram = pentagon plus five triangles", what this refers to cannot be seen. It states a rule. And of course the rule may have been suggested by seeing it thus. This shows what role may be played in a demonstration by attending to an aspect. As mentioned before, Sheffer might have called Russell's attention to the fact about the formula $\sim[\sim p\cdot\sim\{\sim r\cdot\sim q\}]\cdot\sim[\sim q\cdot\sim r]$ that one logical constant could be used instead of two. This is an aspect one might not see even though the formulae of *Principia* always used "~" and "." in this kind of order. To see this is to see another aspect. And looking at a symbolism under a different aspect comes to changing the symbolism. But there is a great temptation to suppose that when one gets a new experience that #this experience teaches one something about the essence, the internal nature, of the formula. It seems to teach one a mathematical (or logical) truth, and this does not seem to be a rule of grammar but a truth about the nature of things.#

The statement that the picture puzzle consists of a man plus other lines could be two different things, a description of what was actually seen (where one's activity of attention is on no different level from an activity with chalk), or a statement that the *picture* was to be so used as to mean "a man plus other lines". The great temptation is to say that if the puzzle were not of this kind it could not be divided into a man plus other lines, or that | | | | | | *could not* be divided into 3 and 3 if it did not consist of these parts. This language suggests that there is an obstacle preventing a division, say, into 3 and 4, that such a division would present an insuperable difficulty. But there is no insuperable difficulty. It just means nothing to say it is divided into 3 and 4. The person who says I cannot divide it in this way must explain what it is like to do it. But naturally he fails since he himself does not admit of the description "dividing into 3 and 4" except in the case where one starts with 7.

Suppose a person divides 1 by 3 to see whether 4 turns up in the development. I tell him "You will never get 4; it is hopeless", and draw his attention to the fact that the dividend and remainder are the same. This may never have struck him. Here it looks as though by drawing attention to this fact we did not perform an operation but showed what was already there. It also looks as though this is a quick way of showing what could be shown by carrying on the division to an enormous number of places and concluding that it it is hopeless to look for a 4. The use of the rule to show by a short cut that a 4 cannot be found looks very like giving up looking for a pine tree on being told that pines never grow in the soil of that neighborhood. But there is a very great difference. Nor is carrying on division to an enormous number of places analogous to looking through a telescope and seeing a long row of 3's. Whatever the apparent analogy, it is mistaken. We do not have here another method of seeing the same thing. However, there are two ways of answering the question whether at the 50th place there is a 3: by writing out 50 places, and by looking at the division 1:3, seeing
$\underline{1}$
that dividend and remainder are the same, and thus anticipating the long calculation.

#Imagine this operation: construction of a decimal by multiplying again and again .25 × .25: .625625625 . . . And consider the order: Look for an 8 in it! What is it like to try to find an 8? What is it like to find an 8? I can hope to find an 8 in the product 284 × 379, but not in this decimal. To say it is hopeless to find a certain result *really* means: *our calculation has already shown it to be wrong* or, our calculation has already *decided against it.*#

Lecture VIII

Notes of this lecture and subsequent lectures by Margaret Macdonald, together with preparatory notes by Wittgenstein

We said that it could be shown that a 4 would never appear in the division of 1 by 3, by calling attention to the fact that the remainder was always 1, and therefore that the result would always be the same. Consider the division of 1 by 7. *#In 1:7 gibt es ein endliches Problem und ein unendliches.#* Now it might not be noticed that there comes a point in the division where the remainder is 1 and the digits of the quotient repeat. 7) 1.0000000 (142857

```
30
 20
  60
   40
    50
     10
```

It seems as if we could make a prophecy that since the series recurs it is hopeless to look for a 6, that there could not be a 6. Similarly, suppose someone were looking for a product whose middle digit was 4, and I said, "Multiply 19 by 34". 19 What would it be like to look for

```
 19
 34
 76
57
646
```

a 5 in this result? I can imagine what it would be like to find a £5 note in a book, but can I imagine what it would be like to find a 5 in the result of this multiplication? I could rub out 4 and write 5; but this won't do. You could object that I have *written* 5 but not *found* 5. To find a 2 in the division of 1 by 7 you might say is easy: here it is. But inasmuch as you know what it is like to find one 2, it looks as though you also know what it is like to find two 2's. "Finding", however, should mean finding by correct calculation. A person who found an object in a maze by climbing a tree and overlooking the maze would be said not to be playing the game, that finding it means going into the maze and searching for it. Similarly, by multiplying 19 × 34 I have shown you what it means to find a 4 in the product: finding it by a process of multiplication, i.e., by a particular operation and in no other way. To describe what it would be like to get two 2's in the quotient of 1:7 you must always describe a faulty calculation, so that it is not finding a second 2. By "find" is meant, find by correct calculation.

Suppose the problem was to find the number occurring at the 10^{10}th place in the quotient of 1:7. One person might try to calculate the number, another give the rule for finding it. #You can now say, it seems, what the 10^{10}th place *will* be. How can one calculation anticipate the result of another? What does it mean: to prophesy what one will *correctly* find?# Are we to say that the two calculations must lead to the same result? What we do say is that if we do not arrive at the same result in both cases a mistake has been made in one or the other.

What is the connection between the meaning of a statement and its verification? Suppose I asked what it is like to find a man in the next room. You might say it consists in going in and seeing him. You have given a description of what it is like. This may mean that a definition has been given of what it means to find him, or it may only be that a connection has been made between one sentence you utter and another. It is not a definition inasmuch as there may be other ways of finding whether there is a man in the next room. What I am saying is that I am describing what it is like for *p* to be true by giving a grammatical connection between *p* and other propositions. It is not necessary for there to be such a connection, but if I give a connection I am saying what it is like for *p* to be true. I am saying something about the grammar of the proposition. To give the grammar of the proposition *p* is to give the sentences with which "*p*" hangs together and to say in what way they are connected.

Lecture IX

We are constantly misled by having the same forms of expression for mathematical and empirical facts. We say, for example, that one rod is longer than another and also that 6 feet is longer than 5 feet. We talk about finding out the same fact in different ways, and of finding the same mathematical result in different ways. But these are utterly different. Matters of fact always involve time; mathematical facts or propositions do not.

In this diagram of 12 dashes it is shown that in these three fours there are four threes, that two processes lead to the same result. What would it be like not to get the same result? We say that we cannot imagine any result except the correct one, nor that the different ways should lead to different results, and this is not because of human incapacity. I could imagine a totally different result, but not in the same game.

We can see what it is like to get a 2 in the development of 1:7. In a sense we know what it is like not to get a 2—simply not writing it down or not doing a calculation at all. But this is quite different from getting a 6 instead. If I say I cannot imagine a 6 in the result, this means that the calculation shows me what it means to imagine a 2 and gives no sense to the statement "I imagine a 6 in the result". The same thing applies to getting the same result by two calculations. That I cannot imagine their not leading to the same result means that the proof of their leading to the same result shows what it is like for them to do so. The division into threes and fours does not show two processes leading to the same result but rather what the result of two processes is like when they lead to the same result. I might say that the following diagram shows what we call two things meeting; but it does not show two things meeting: ——→←——

There are many ways by which it might be made possible for me to look into the next room, by breaking the wall down, blasting it with dynamite, making it transparent, unlocking the door. None of these ways is part of what is meant by achieving the result. On the other hand, 10^{10} is defined as a number got at in a certain way. And there might be other ways which lead to the same result. For example, a slave might be got to write down 10^{10} digit by digit, and another man might get it immediately by the formula. This looks like doing the same thing a quick way and a long way. It looks as if it were in the *nature* of the processes to lead to the same result. But the processes themselves show nothing about a *must*.

The problem of finding different ways of reaching the same result in mathematics may seem analogous to finding different ways of looking into the next room. It might be said that any way of solving the problem will do, so long as it is in accordance with the rules of arithmetic. But the ways will differ according to the system of arithmetic. What one calls mathematical problems may be utterly different. There are the problems one gives a child, e.g., for which it gets an answer according to the rules it has been taught. But there are also those to which the mathematician tries to find an answer which are stated without a method of solution. They are like the problem set by the king in the fairy tale who told the princess to come neither naked nor dressed, and she came wearing fish net. That might have been called not naked and yet not dressed either. He did not really know what he wanted her to do, but when she came thus he was forced to accept it. The problem was of the form, Do something which I shall be inclined to call neither naked nor dressed. It is the same with a mathematical problem. Do something which I shall be inclined to accept as a solution, though I do not know now what it will be like.

Lecture X

It looks as if one calculation, by a given rule, tells one the result which another calculation, the actual working out to a thousand places, *must* have. Is there a difference between "there *must* be" and "there *is*"? The "must" always refers to what may roughly be called a method and a calculation not yet made. This suggests that there is something known beforehand. But you might say beforehenad that an 8 would appear in the sixth place and something else might result. What is meant by the result of a process? Compare the result of heating water vapor to a certain temperature with the result of moving a piece of chalk until it ends up where one stops. In the first case the process can be described without describing the experiential result of splitting into hydrogen and oxygen. In the second, where the result of the process, ending up here, is part of the process, describing the process includes describing the result. "Result" is used in two different ways.

A mathematical process is not such that it could be what it is and the result be a different one. To say a process gives a certain result means giving the result. #In one sense you can't know the process without knowing the result, as the result is the *end* of the process. A calculation leads to a result mathematically apart from whether it has actually been performed. In another sense you can know a process and not know the result. "In what sense is it possible not to know where a *mathematical* process leads? We could answer, It is possible not to know where it *will* lead, but not possible not to know where it leads".*#

Suppose a man worked out a multiplication once and then when he needed the product again, worked it out afresh. It might be said that he had done it once and is bound to get the same result. Can you imagine a man not seeing this and so having to get the result every time anew. Suppose someone counts the five dashes | | | | | first from one end and then from the other. Must he always get the same result? There is an

*Editor's translation of Wittgenstein's note: *In welchem Sinne ist es möglich nicht zu wissen wohin ein* mathematischer *Vorgang führt? Man könnte antworten, es ist möglich nicht zu wissen wohin er führen* wird *aber nicht, nicht zu wissen wohin er führt.*

empirical question involved here, but also something not empirical. In the visual five we correlated each dash with a numeral: 1, 2, 3, 4, 5. Must the series give the same number whichever way it is counted? What sort of fact is it that the order does not change the result? How could it make any difference if he began at the other end? He might agree that it could not, or again, he might not agree. Suppose someone asked, How do you know that you have not left out a digit when you counted 1, 2, 3, 4, 5? You might reply, How could one discover such a mistake?

The first man who discovered periodicity in 1:7 found a way of discovering in the first 1000 places whether there is a 6. Suppose that at first he wished to go through the first 1000 places and then discovered periodicity, whereupon he changed his mind about the problem. Had his problem been to find out whether *he* would write a 6 in the first 1000 places, he could solve it only be seeing what he would in fact discover by calculating. The second method, involving the discovery of periodicity, would not help him solve that problem. But it would determine what it was *correct* for him to discover.

One might say that it can be *proved* by induction that the two methods lead to the same result. A proof by induction is such that you can always say, It must go on this way. We must distinguish between seeing the calculations actually written down and seeing that it must always go on this way. Must one recognize periodicity as a proof that there will be no 6 in the development of 1:7? No. There is no reason why a person should see that it must go on that way. He may accept this as a proof for future occasions, or he may not. We cannot make him recognize a new proof. In a way it is a matter of experience that everyone will find the same number by calculating 1:7, though it is an important fact of experience. That in mathematics we have to recognize that two methods must give the same result is not a fact of experience but a rule. But we accept the rule because we find that in all calculations we do get the same result.

Can there be a 4 in the development of 1:3? The way to find out would be to divide. But how long is one to go on? Without knowing this, one has not been asked anything. It then looks as if by finding the formula, and seeing that there could not be a 4, one has found a short-cut to infinity. What happens is that one accepts the formula as an interpretation of the question and of the answer. A proof is not found but constructed. Periodicity does not mean the same as several repetitions of the same number or numbers, but makes a new calculus between the dividend and certain remainders. Without this calculus division was not in any way incomplete. When you accept periodicity you accept a new interpretation of your question and a new method of answering it; but it looks as if you *must* accept the periodic result unless you are a fool.

Lecture XI

Consider |ıııı|ıııı|ıııı|, and the question, "How many times can one count four dashes between the strokes?" The answer is three. About this one can say two things: that one gets three and that one *must* get three. There seem to be two independent processes which lead to the same result, one that finds three, and the other that fixes what the result must be. Finding three seems to be some kind of mathematical experience. If you put a bit of string between two poles and say it fits, realize that it fits because you made it fit. We have here the picture of what it is like for something to fit something. Similarly for a hollow mould or cylinder for which one has a hard object made to fit it, or a lump of clay which one could make fit it. In all cases in which two proofs meet it looks as if one has two independent processes leading to the same result, and not as if one had made them do so. By one process it looks as though we find an end and by the other reach it; and it looks as though we ought to be surprised at reaching it.

In a chess game two players might be said not to see a simple truth with regard to a move, say, that if a bishop is moved to a certain square the result will be a checkmate. But what truth don't they see? Not a truth about pieces of wood; moves with these might be regarded as experiments, in which case the result could not be predicted. The pieces might break, burn, etc. What we mean is that they do not see that the process must always lead to this result. There is a difference between a process having a result and being its own result.

We might say that if two processes *must* lead to the same result, showing that they do is a confirmation. But whether it is a confirmation or not may be doubted. #After you have seen that 1000:3 must lead to 333, is it a confirmation to calculate it and see that it does? What does it mean to say that one calculation confirms the result of another?#

In the division of 1 by 7 to thirty places, we get, say, five periods. That one actually writes down a certain number of repetitions is a fact of nature. But we also say that we must get five. Should we call it a mathematical coincidence that when we write down the calculation five repetitions have occurred after thirty places?* Or that a 5 must occur at the 17th place? The proof that 1:7 must lead to a 5 in the 17th place can be put beside two different things (1) the actual written row of digits, (2) what may be called the method for producing the 17th digit. The latter can be interpreted in all sorts of ways, though we may not as a matter of fact interpret it in all sorts of ways. To say of the many methods which might be shown you as leading to the same result, that they *must* lead to the same result, looks like a prophecy; but really it is a resolution we have made. That they do lead to the same result is a physical fact. The resolution did not prophesy it, but we operated in accordance with the resolution we made.

If we were asked, we would probably all say that words like "great", "small", "hot", "cold" are relative and not absolute terms. But in ordinary life there is a use of such words which is absolute. As sensations, *hot* is not a high degree of *cold,* nor *cold* a low degree of *hot.* And it is the same with *agreeable* and *disagreeable* and with *great* and *small*. We always think of the infinite as something very huge or very tiny. The idea of converging to a point is the idea of convergence on something infinitely small. But the infinite has nothing to do with size at all. There is a constant temptation to picture an enormous extension when we find the remainder in a division equal to the dividend. We take this to be the criterion for infinite periodicity, and say the result can be infinitely repeated. And it looks as if some superhuman being might survey the infinite extension even though we cannot. The greatest puzzle is that in some queer way what has not been done, say, division to the 17th place, seems as though it has been done, as though the whole extension has been given.† We tend to think of the development as an actual enumeration. If this picture of an

* *#Gibt es einen Zufall in der Mathematik?#* Taken from Wittgenstein's notes for this class lecture. (Editor)

† The following, from The Yellow Book, reinforces this point:

Given the series 1, $1+a$, $1+2a$, . . . , where "and so on" trails off into saying silently four additional numbers, what one has in this case is a series that comes to an end: three numbers said aloud, followed by four numbers trailing off. This idea of "trailing off" makes us fail to realize the difference in grammar for "and so on". It creates the illusion that we have done the counting and not done it. We behave as though the numbers which trailed off were fixed, that they were *meant* though unsaid. But have you meant them if they are unsaid? I am only saying that a number not written down is not written down. We in fact have three numbers and "and so on", each with their own grammar.

For discussion of the use of "and so on" to indicate endlessness, see *Philosophische Grammatik*, Part II, Section II, 10. (Editor) enormous extension is given up we see that the infinite is on a totally different level [from the finite]. The difference is like that between a race to a goal and an endurance race without a goal. I do not mean by this that the infinite is "unreal". The word "infinite" has its uses. For example, to say that there is no 6 in the infinite development of 1:7 means that it is not in the period, and that is all. It only makes sense in this calculus. How is it that this has been so misunderstood?

Lecture XII

Are there three consecutive 7's in the infinite development of π? There is in fact no means of showing that there must be or cannot be three 7's in the infinite development. It is queer that we can ask this question and have no means of discovering the answer.

In talking about the infinite development of π we use the picture of something developing, of its growing longer and longer, stretching out ever so far. We talk of an unlimited choice, as compared with a huge limited choice, as though it was something of the same kind, only more huge. We think that thousands of millions is nearer to infinity [than a thousand]. But infinity has nothing to do with size.

If I ask, Is there an 8 in the infinite development of 35:161?, you will develop to a period and if 8 appears you will answer, Yes. What you know is that it has appeared in the period, or has not, as the case may be. This is different from asking the same question if you do not calculate with periodicity. The question is then equivalent to: Find some calculus which will lead you to say whether there is an 8 in the infinite development. We might say the questions are the same, only in one case you have the tool and in the other you have not—as if the order were "Cut this book in half", when in the one case you have a suitable knife and in the other you have not. But it is not quite the same, for there is not a psychological question here corresponding to the physical question connected with cutting the book. We must examine sentences in which we use the term "infinite development", to find out what meaning we have given them. What we call the infinite development is bound up with our method of producing it.

We could say that since we have made up π, we have made up all its consequences. Similarly, we might say that after we have accepted certain axioms logic compels us to go on in a certain way. Now what sort of compulsion is this?

We use "2×2" in such a way that it means the same as "4".

Hence to say "He ate the number of apples which is the product of 2 and 2" is just to say he ate 4 apples, and to say "He ate the square root of 4 apples" is just to say he ate 2 apples. Compare these statements about 4, 2×2, and 2, with "Mr. Wisdom is sitting in this chair" and "The Sidgwick Lecturer in Moral Science is sitting in this chair". These two sentences do not have the same meaning, though they happen to be about the same person. But 4 does not *happen* to be the second place in the development of π; 4 *is* the second place of π. There is no such thing in mathematics as a description of something and its name. That is, there is no such thing as the product of 35 and 45 and the number 1575 which happens to be the number described; they are the same number. In this example we have another way of bringing out the fact that a process in mathematics contains its result.

Suppose someone asks, Is there a 9 in the infinite development of π?, and we calculate and find a 9 in the fifth place. You might then say that since there is a 9 in the fifth place there must be one in the infinite development. Now if you like to make that rule there is nothing wrong with it. But what have you said? Does your saying that you are justified in asserting there is a 9 in the infinite development of π entitle you to say there is some specific number, say 8, in the infinite development of 1:7? if you have given a meaning in the one context [to "an x in the infinite development"], you have not necessarily given it in all. You have not been asked anything until you have been told what to do.

Lecture XIII

If one asks whether there is a 7 in the infinite development of a number, periodicity will give one an answer. Suppose you are calculating with periods and define "There is a 7 in the infinite development" as "there is a 7 in the period". That would be useful. But if you say there is a 7 in the period and therefore in the infinite development, you are concluding from a useful phrase to a verbal phrase without meaning. There are two criteria for there being a 7 in an infinite development (1) finding a 7, (2) finding a 7 in the period. If the number is to be compared [as to magnitude] with a rational number, one must also have a criterion for showing where in the development of a rational number the period will end. If there is no means of determining the length of the period, then to say a number is periodic changes the sense of "periodic" entirely. Suppose someone claimed to find a number periodic although he could not say when the repetition began. He would say it must be a rational number. But *which* rational number?

Suppose we have been taught to develop root 2 so as to get a series of decimals: 1.4, 1.41, 1.414, 1.4142... Would you in that case have reason for calling the square root of 2 a number? You would be more inclined to say it is a rule for developing these decimals, and you could talk then of giving root 2 an index, say 5: $\underset{5}{\sqrt{2}}$. We talk of root 2 as a number because we can construct a method for finding whether it is greater or smaller than any given rational number.

We talk as if we have a series of infinite length corresponding to the numerical symbol $\sqrt{2}$. Now $\sqrt{2}$ is a symbol for which there are rules for indefinite development. Suppose I give you a new irrational number, symbolized by $\frac{7\rightarrow0}{\sqrt{2}}$, constructed by putting 0 whenever 7 is reached in the development of root 2. Are there any objections to calling this a real number? It might be said that I do not know whether I shall ever get a 7, or it might be proved that there is no 7, in which case $\frac{7\rightarrow0}{\sqrt{2}}$ will be $\sqrt{2}$. There is the idea that there is a development corresponding to $\sqrt{2}$ even if we do not happen to know it. But before $\sqrt{2}$ had been calculated we did not have a connection of $\sqrt{2}$ with 1.414. What we have are rules of such a sort that a calculus of greater and smaller between irrational numbers can be made up. The order, "Replace 7 by 0" has no sense except that it gives a rule. The rule does not tell one how long to go on before replacing 7 by 0. There is nothing to know about root 2 except what we have laid down. It is not that there is something to know that we don't know, but rather that there is no calculus yet for it.

To say a certain rational number has a period that we do not know is to use words in an entirely different way. It might be said that if we have a proof that there is a period then there *must* be one. I want to say that it is most misleading to say this. There is a very loose relation between a mathematical proof and the words in which the result is stated. The relation between the proof and the words which express what the proof proves is widely different for different kinds of proof. Proof that there is a 7 in the infinite development of π may mean all sorts of things. It is easy to succumb to the old simple absurdity that there is one development which is *the* infinite development of a number and that our problem is to find an indirect method of knowing something which the infinite Being or God knows already in its entirety. (Compare Russell's misleading claim that we have no direct acquaintance with an infinite series but have knowledge by description of it.)

The relation between the proof and the English (or German, etc.) words in which the result of the calculation is expressed is very different for different proofs. We say there is a proof that $26 \times 13 = 338$, and a connection between this and the proof that $m \times n = r$, where m, n, and r are replaced by any numbers. You know what this proof is, viz., multiplication. But if you say that proof that the trisection of an angle with ruler and compasses is impossible, you do not know what that proof is, in the sense of knowing whether it belongs to a system of proofs. You might say it belongs to the class of proofs of the impossibility of so-and-so. But this is a very rough description and quite different from the multiplication proofs. If someone has proved that $26 \times 13 = 338$, one knows what he has done. But if I say there is a proof that a period exists but that I do not know where the period begins, I do not know what sort of proof it is. In the case of a chemical process, to know what the process yields one does not look at the process but at the result. But with a mathematical proof one must look at the proof to know what it yields. The formula of the proof may or may not give a catalogue of proofs. In the case of such things as multiplication and division it does.

You can imagine the result of a proof as being like the end surface of a body. Suppose we had cylinders of a certain width and length, all of them being of the same length and different widths. We could give a catalogue of them by means of their end surfaces, and we could find the volume of any one by looking at its end surface. But were the lengths to vary we could not catalogue them by means of their end surfaces alone. If for a variation in width they varied an inch in length we could again catalogue them by their end surfaces. We could call the result the end surface of a proof, and the proof the body. All proofs such as $26 \times 13 = 338$ could be classified by their end surfaces, whereas the end surface, "It is not possible to trisect an angle with ruler and compasses", does not help to catalogue the proof at all. We could classify this proof with other proofs such as the proof of the impossibility of constructing a heptagon. Then the result of the proof may help us to say what the proof is. When this occurs I say that the result of the proof is bound up in a mathematical way with the proof.

Otherwise it is a prose sentence so loosely connected with the proof that lots of other prose sentences might be connected with it.

Certain verbal forms are misleading for a time and then cease to be misleading, for example, symbols for imaginary numbers. There may have been something misleading in the idea of imaginary numbers when it was first introduced, but now it is utterly harmless. It does not mislead anyone. By contrast, to say the appellation "infinite development" is misleading is correct, though of course it does not mislead anyone in his calculation. It misleads people into a wrong idea about what they have done. The idea of the infinite as something huge does fascinate some people, and their interest is due solely to that association, though they probably would not admit it. But that has nothing to do with their calculations. I might say that chess would never have been invented apart from the board, figures, etc. and perhaps apart from the connection with troops in battle. No one would have dreamed of inventing the game as played with pencil and paper, by description of the moves, without the board and chess pieces. Still, the game could be played either way. It is the same with mathematics. It is the associations of the calculus which make the calculus seem worthwhile. But these are quite different from the calculus. Sometimes the associations are connected with practical applications, sometimes not. But without the associations with the huge, no one would care a damn about the infinite.

In a recent article in *Mind** the question was raised whether if someone had proved it is not self-contradictory to assume there are three 7's in the development of π we should say that he has thereby proved that there are three 7's, even though the proof gives us no method of finding three 7's in the infinite development. It was suggested that one should distinguish between two expressions, (1) (written in Russell's notation) $(\exists x)\phi x$, i.e., there is a place for which it is true that three 7's begin there, and (2) $\sim(x)\sim\phi x$, i.e., it is not true that for all places there are not three 7's. The author suggested that there are two methods of proof, (1) showing three 7's between two places in the development, (2) showing that it is self-contradictory that there not be three 7's. $(\exists x)\phi x$ (where three 7's have been exhibited) should be given as the result of one proof, and $\sim(x)\sim\phi x$ (meaning that it is self-contradictory that there not be three 7's) as the result of

* "Finitism in Mathematics I", *Mind*, XLIV, no. 174, by Alice Ambrose. (Editor)

the other. But this is no solution. It is like saying to two people who are quarreling about a book, "See, this will settle your quarrel: let one of you take the title and the other the rest of the book." Of course that would not satisfy them, for the title just is part of the book, it belongs to the book. It makes a division where people would say there cannot be a division. I do not mean that we cannot divide it like this; only if we do so there is no earthly reason for assuming the ordinary notation for (2). The two are considered inseparable. If I say the one proof proves (1), so does it prove (2). If I do not wish to say they prove the same thing, there is no use making the division there. To separate the two in the suggested way is like dividing a double house by giving one house and the kitchen to one person and the dining room and the rest of the house to the other. This would be said to be inadmissible because the kitchen and dining room go together. There is no use in separating where normal language revolts against the separation. If I said, "Divide the double house so that one person has one house and the other has the other house", that would be different. We might say that whoever possesses one of these expressions thereby possesses the other, and that the separation must be in a different way. We must then state what reason we have for saying that one proof proves the existence of so-and-so.

Lecture XIV

From what we have said it seems that a mathematical statement has no sense before being proved true or false. In the case of the statement that there are three 7's in π there is no proof that it is self-contradictory nor do we know that there are three 7's. It seems as though asking the question is senseless since we have no means of answering it. Suppose, however, that we discover three 7's or devise a proof that it is self-contradictory that there should be three 7's. The question seems both to have been answered and to make sense. This account of the possible answers is contrary to what we ordinarily call a proposition. For we say a proposition must make sense before we know whether it is true, or false.

Proceeding from the analogy between a finite extension and the infinite development of π, we are tempted to go on to say the same things about both, in this case that there either are or are not three 7's in π. This is the sort of thing some logicians would say—there either are or are not. But why do they repeat the law of excluded middle? What does it say? It is a tautology. Why do they stress this, and not, for example, the law of contradiction? They do it in order to conjure up a particular image—as it were, of something lost in infinity. Consider the fact that in leading a dog, the longer the leash the more freedom for the dog. Now suppose I say the lead is infinitely long. Then I might as well say I do not lead him at all. Analogously, if I ask, "Are there three 7's in this infinite series?", I might as well say the question cancels itself. Its grammar is such that it is not a question.

One can look at this question in two distinct ways. (1) If thought of in terms of an extension, the question whether there are three 7's in the infinite development of π will be thought to make sense, because it makes sense to ask whether there are three 7's in any finite development, the difference between the finite and the infinite being only a matter of degree. Just as the third place of π is 4 whether we know it or not, so some places of π are (or are not) 777 whether we know it or not. Those who concentrate on the extension of π say the question makes sense, whatever means we have of answering it, or even if we have no means at all. They imagine a method which would give us the result if only we could apply it. It does not matter that we have to resort to a series of dodges to get some approximation to the result, for this is only because we are finite creatures and the difficulty is purely psychological.

There is another way of looking at the question, and this leads to a different difficulty. According to this point of view, (2), it makes no sense to say there are three 7's in π, or to behave as though there were some method of finding out though it is impracticable. No method has been fixed, just as no goal has been fixed in an endurance race. Therefore the only criterion for there being, or not being, three 7's in π is the proof after all. On the one view there is a way of finding out, only it is utterly impracticable for us. But it exists, and gives the question sense. It is only that we have an indirect way of getting at the same result the Deity could get at directly. The other view says, No, this claim about the Deity seeing the whole extension of π means nothing, and the only criterion for there being, or not being, three 7's in π is the actual proof, if there is a proof. The intensional view (2) is that either one has a proof that there exist three 7's in π or one has a proof that there cannot be three 7's in π. There seems to be still a third alternative, that one has no proof one way or the other. When Brouwer says that the law of excluded middle does not always hold he is taking the intensional point of view.

But this view creates another difficulty. It now seems as though the question had no sense. By contrast with the first view, which stresses the point that what we call a proposition is something true or false regardless of whether we know which it is, the second view is the difficult one that there seem to be propositions which have no sense until we know whether they are true or false. The difficulty can be put in this way: wherever there is a proposition, we say, there ought to be a question, e.g., "The man is black" and "Is the man black?" And in "Are there three 7's in π?" we seem to have something we should call a question. The difficulty is that this question must be answered in order to be called a question. But is not this characteristic of all mathematical propositions? It is, for this reason: if one takes any proposition at all, say that $26 \times 13 = 419$, one can say that the result cannot be imagined to be 419 if it is not 419 and that it cannot be imagined to be other than 419 if it is 419. This shows straightaway that mathematical propositions are different from what we ordinarily call propositions.

The trouble into which we fall on the intensional view is due to one very simple matter of fact: that what we call a proposition in mathematics, and what we call a question, can be all sorts of utterly different things. For example, the proposition $26 \times 13 = 419$ is essentially one of a system of propositions (the system given in the formula $a \times b = c$), and the corresponding question one of a system of questions. The question whether 26×13 equals 419 is bound up with one particular *general method* by means of which it is answered. Let us compare the proposition which is its answer with one that is totally different, the fundamental law of algebra, viz., that every equation has a solution. This has the form of a proposition and is written as an ordinary English sentence. But it is in a totally different position from the multiplication proposition. It seems to be an isolated proposition, unlike the latter. Also, it seems to get its sense from the proof, while the propositions stating what the product in a multiplication is do not. Whatever the answer to the question, "Has every equation a solution?", nothing more would be said by it than what the proof gives. By means of what we call the proof of there being roots to an equation we really know what proposition has been proved, and we know the answer to the question. Would you understand me if I said that the answer here has much more in it than the question did? Normally it is not like this.

In the case of the question about the product of 26 and 13, there is something about it which makes it look like an empirical question.

Suppose I ask whether there is a man in the garden. I could describe beforehand a complicated way of finding out whether there is or not. There is a resemblance of the multiplication question to this one, in that before you find out I could tell you how to find out. But when we ask, Does every algebraic equation have a root?, the question has hardly any content. It gives us a sort of hint as to what we are to do, but the proof provides it with its content. So the proposition which is its answer is of a totally different kind from a proposition of the form $a \times b = c$. The reason I have brought up this comparison is to see what sort of proposition "There are three 7's in π" is.

Suppose someone asks whether all algebraic equations have roots, and defines a root as a real or complex number which when substituted in a given equation makes the two sides equal. The question has obvious sense if you like to define it that way. Presumably all the numbers are there and the Deity could try them all to see whether any of them produced an identity. If the hypothesis of a mathematical Deity is discarded, then we must answer the question, What is the criterion for every equation having a root? Could we say *what* would be proved? The difficulty here is the same as in the case of proving that there are, or are not, three 7's in π. We do not know at all what the proof would be like. In this respect the questions as to π and as to whether every equation has roots are alike, and they are unlike such questions as "What is the result of 26×13?", "Is there a 4 in the product?", "Is there a 7 in the period of 1:7?", "Is there a 7 in 1:27?", which can be answered by Yes, or No. These latter belong to a whole system of questions. We have a method of answering them, and the answers within the system of answers are like ordinary empirical propositions in the respect that one could give a method for deciding them. If we had a method not merely for answering whether there are three 7's in π but also whether and how often any given group, say 1,9,5,6, occurred, then the question about π would roughly be of the same kind as the question about multiplication. The proposition would be more or less like what we ordinarily call a proposition. On the other hand, had we something which claimed to prove nothing more than that there are three 7's, without showing where they occur, we should not have a proposition comparable with others of a system, and it is doubtful whether we should call it a proof. One might say that the verbal expressions in mathematics which we use to describe the results of proofs are used highly metaphorically. They only get their strict sense from a method, and when the method has been evolved, then questions in that system become very like ordinary empirical questions.

Compare the question "Is there a 4 in the product of 26×13?" with the question "Is there a 4 in the first 100 digits of π?" Here we have two questions formally similar but at the same time different. Now suppose that we first develop 100 places of π and if after that we find three 7's we replace the first digit of π by 4. Then ask, "Is there a 4 in the first 100 places?" We now have no idea what to do to answer the question.

What we call a proposition is not one thing, aloof and isolated. When people ask whether the question, "Are there three 7's in π?" is sense or nonsense, they are up against the difficulty of saying under what conditions we would call "There are three 7's" a proposition. How far is it like the multiplication case and how far like the fundamental law of algebra? What is in the normal sense a question, or proposition, becomes by isolation something which loses every character which it appears to have as question or proposition. "Are there three 7's in π?" belongs to a huge system of questions, but only the proof will show to what system it will belong. Only so far as the proof is a member of a system of proofs is the English sentence expressing the result of the proof justified.

Lecture XV

We said that if one took the intensional view, a question or proposition does not make sense until a proof, or method of proof, is given. We may say there is a contradiction between this use of "proposition" and its use in ordinary life. Everyone talks more about eggs and bacon and tables than about mathematics, and the use of the word "proposition" for statements about such things is the one to which we have been trained and to which we are accustomed. Now some of the usages of "proposition" in mathematics do not go against the most common usage at all. We normally use the word in such a way that for any proposition there is a question bound up with it. If the proof is what gives sense to the question, this seems to contradict what we mean by a proposition.

The cases in which a mathematical question is similar to an ordinary one are those in which we have a general method for answering it. For example, since we have a general way of deciding whether $m \times n = r$, the question whether $26 \times 13 = 1560$ resembles an ordinary question, although it does differ from an ordinary question since we cannot imagine what it would be like for the answer to be true if it were false, or false if it were true. One might say that in mathematics the idea of a question is bound up with the idea of a mistake in calculation. Suppose I asked, Is $26 \times 13 = \sin a$? Is this a question or a mistake? As a rule we should not call it a mistake, for we limit what we call mistakes to a few things. We should be inclined to call it nonsense. Whether an expression has sense depends upon the calculus. I can imagine the kind of mistake which would lead one to say $26 \times 13 = 1560$, or that 4 is the first digit of π, and thus I could say that the corresponding questions about them are genuine.

To say that it is the proof which gives sense to the question seems to contradict what we ordinarily mean by a question. But note that after discovery of the period in the development of 1:7, the question whether there is a 4 in the infinite development no longer poses the problem of how we are to develop it far enough to find out. The question about the infinite development becomes utterly unimportant. It is as though the *stress* of the question has moved away from the word "infinite". Periodicity is part of the method, and the period when found is involved in the question as well as in the answer.

To say that it is the proof which gives sense to the question is absurd because it misuses the word "question". But it is not absurd to say that it is the proof which gives a method of answering the question and in this way gives sense to the question. The question is thereby embodied in a system of questions corresponding to which there is a system of answers.

It is often said, Suppose there were a proof that so-and-so. This kind of supposition says not one jot more than is said about the proof. It is like saying, Suppose there were a system in which $26 \times 13 = \sin a$. In saying this I do not point to such a system; I merely write down "$26 \times 13 = \sin a$". Mathematicians talk of the possibility of there being a hidden contradiction in a proof. But in supposing it possible to find a contradiction they have not supposed any more than they have written down. Similarly with Ramsey's supposing that there were a universe which contained only three individuals. This is not at all like imagining what the room would be like if it contained three chairs instead of six, or what the earth would be like if there were only one town on it. What is imagined is what he writes down and nothing more. He has said nothing about what its use might be. One cannot conclude anything at all about what the world would be like if it consisted of three individuals.

Consider now the supposition that there is a proof of three 7's in the infinite development of π. You might remark, Well, if there were, I suppose it would be a proof like so-and-so. In reply I would say that there was nothing behind the supposition, but only something before it, namely, your use of it. It may be that psychologically we are more ready to make one kind of connection than another, but until it is made we are not driven to making a connection in any one particular way. Words like "Suppose there were a proof that so-and-so" get their justification only in terms of what is done after they are said. Suppose I claimed to imagine there were not three 7's in π. You might say I was imagining π divided into chunks of digits, with the proof proceeding by induction. What have I imagined? I seem to say something about a proof of there not being three 7's in π, and that is queer because I have not got a proof. I have done nothing but operate with the expression "proof that there are not three 7's".

To make a supposition would normally be to have some sort of picture of the kind of thing that is being supposed. If I suppose that this room is higher than it is, I might have a picture that represents how the room would look, and other things that are consequences of the supposition, e.g., that it would be more difficult to heat. But if I say, "Suppose I have a proof . . .", I have nothing but those words. What comes after that, what I then say, is all that the supposition consists of. The builder to whom I say, "Make me a room one foot higher than this", knows exactly what to do; and his knowing what is to be done may consist in having a drawing of the room before him. With the supposition about π I have no drawing and cannot supply one. And it is essential that I should not be able to supply one. This shows that what we call a supposition in mathematics is entirely different from what we call a supposition in ordinary life. There is the same contradiction between the uses of "supposition" in the two contexts as between the uses of the words "proposition" and "proof".

Suppose someone says he has found a proof that there cannot be three 7's in π, and that someone else claims he has found three 7's. Is the proof going to show the latter where he made a mistake? Or is it not to make any difference at all? If you say it need not show him where the mistake is, then it will be utterly different from anything we call a proof in ordinary English.

PART IV

Philosophy for Mathematicians

Wittgenstein's Lectures

1932–33

From the notes of Alice Ambrose

Philosophy for Mathematicians

1932–33

1 Is there a substratum on which mathematics rests? Is logic the foundation of mathematics? In my view mathematical logic is simply part of mathematics. Russell's calculus is not fundamental; it is just another calculus. There is nothing wrong with a science before the foundations are laid.

I shall exclude all questions that may be solved by luck or experience.

Consider the question, What *is* the number 2?, and the definition of number as a predicate of a predicate. Now there are all sorts of predicates, and 2 is an attribute of a predicate, not of a physical complex. What Russell has said about number is inadequate, first because criteria for his use of identity are not mentioned in *Principia,* and secondly because the notation for generality is confusing. This notation is built up after the analogy of subject-predicate propositions in ordinary language, such as those describing physical objects. The "x" in "$(\exists x)fx$" stands for a *thing,* a substrate; and propositions having different grammars, both mathematical and nonmathematical propositions, are dealt with in the same way, e.g., "All men are mortal," "All men in this room have hats," "All rational numbers are comparable in respect of magnitude."

We use numbers in connection with many different predicates. Russell said 3 is the property common to all triads. What is meant by saying number is a property of a class? Is it a property of *ABC* (the class), or of the adjective characterizing the class? There is no sense in saying *ABC* is three; this is a tautology and says nothing at all when the class is given in extension. But there is sense in saying that there are three people in the room. Number is an attribute of a function defining a class; it is not a property of the extension. A function and a list are to be distinguished. Russell was desirous of getting another "entity" besides the list, so he gave a function using identity to define it. Consider the class $\boxed{a}$. Russell gave $x=a$, using identity, as the function defining it. Ordinarily, the replacement of a function by a list (class) is mistaken. We say something different when we talk about a class given in extension and when we talk about a class given by a defining property. Intension and extension are not interchangeable. For example, it is not the same thing to say "I hate the man sitting in the chair" and "I hate Mr. Smith." But it is otherwise in mathematics. In mathematics there is no difference between "the roots of the equation $x^2+2x+1=0$" and the list $\begin{array}{|c|}\hline 1\\ 1\\ \hline\end{array}$, or between "the number satisfying $x+2=4$" and "2." The roots, and 2, are not described in the way the person is who satisfies the description "the man sitting in the chair."

2 A pernicious consequence of the attempt to interchange function and list is in connection with infinite lists. What is the sense of talking of an infinite list, e.g., the list of values of a function of two variables? The phrase "infinite list" has no meaning unless given a meaning entirely different from the ordinary sense of "list." This is not to say some uses of "infinite" are not legitimate. Consider a pendulum attracted to bodies according to a known law. One can calculate the way it swings according as there is a finite, or an infinite, number of attracting bodies. Meaning can thus be given to the statement that it is attracted to an infinite number of bodies. Determining the number of bodies by means of the law is entirely different from counting them. An "infinite number" has an entirely different grammar from "finite number." We need not define "infinite number"; rather, we must say how the term is used.

The difference between "finite number of numbers in the development of π" and "development of π" is like that between a railway train and a railway accident. The two expressions are obviously connected, yet have entirely different meanings. How do we learn these two phrases? To explain "the development of π" one need not write down a single number, whereas to explain "the development of π to seven places" one writes down the numbers. Law and extension are utterly different.

These two spirals are related as larger and smaller bit. But the law for a spiral and one of these are not so related. And to say the series of cardinals is *longer* than 1,2,3,4 is to say something different from saying 1,2,3,4,5 is longer than 1,2,3,4.

3 To return to Russell's definition of number, and the relation of similarity, or 1-1 correlation, which figures in it. Just as one could define the length one foot as the length which stands in a certain relation to the Greenwich foot, so Russell said every triad could be correlated to "the Greenwich triad."

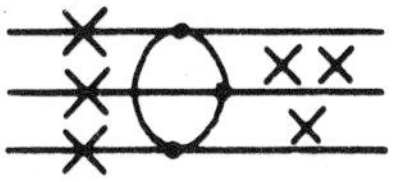

A difficulty in Russell's definition is in the notion of 1-1 correlation. This notion is vague. In his account, correlation is effected by use of the idea of identity. The correlation of A with B is given by the function $x = A \cdot y = B$, as the only things satisfying this logical product are A for x and B for y. There are two meanings of identity in *Principia Mathematica*. One use of the identity sign occurs in definitions, i.e., in a shorthand: $1 + 1 = 2$ *Df*. I shall call a definition a primary equation, that is, an equation one starts with. If $3 + 4 = 4 + 3$ and $3 \times 4 = 4 \times 3$ occur in a calculus in which these commutative laws are definitions*, one sign may be put in place of the other. But what is meant by "$1 + 1 = 1 + 1$?" It is part of the grammar of "=" that one can write this formula. But how is it used? The formula "$a = a$" uses the identity sign in a special way; for one would not say that a may be substituted for a. Yet we do start in inductions with something like $a = a$. Another use of the identity sign occurs in *Principia* in the notation for "There is only one thing which satisfies the function f": $(\exists x):fx\,.\,(y)\,.\,fy \supset (x = y)$. Does it follow from this use that it makes sense to write "$x = x$"?

This use of "=" is confined to cases where an apparent variable occurs, and it could be eliminated by using different symbols for different things. In place of $(\exists x):fx\,.\,(y)\,.\,fy \supset (x = y)$, write $(\exists x)fx$: $\sim(\exists x,y)\,.\,fx\,.\,fy$, which says there is one thing and not two. Russell would write "Only a satisfies f" as $(\exists x)fx\,.\,x = a$. I would write it as $fa\,.\sim(\exists x,y)\,.\,fx\,.\,fy$. Russell's use of the identity sign is the use it has in the nonsensical expressions $a = a$, $a = b$, $(\exists x)\,.\,x = x$. These are degenerate cases of the legitimate use of identity. It is true that $a = a$ and $a = b$ are used at the start of proofs by induction. I suggest not

*Compare these definitions with $13 \times 14 = 14 \times 13$, where we have different calculations on each side of the equation.

using them in induction at all, or else allowing "1 + 1 = 1 + 1" to mean something in this particular game. As for Russell's use of "=", it occurs in the expression of *There is only one man in this room,* which in ordinary English does not refer to a relation $x = y$.

Russell said a and b are identical if they have all their properties in common. It is as if this were a sort of physical law coupled with the feeling that a and b never will have all properties in common. There is a lack of clarity about properties. The properties of physical objects have only an irrelevant similarity to the properties of numbers, straight lines, etc. in mathematics. Property terms in ordinary contexts must stand for qualities that it is sensible to say the substrate *has,* or *hasn't.* It is nonsense to attribute a property to a thing if the thing has been *defined* to have it. Compare the answers to the question "What properties has the color red?": (1) it is a property of red that something has it, (2) it is a property of red that it is darker than pink. There is no such proposition as *Red is darker than pink,* because there is no proposition that negates it. What meaning has "Red is not darker than pink?" To say that red is darker than pink is not to talk of a property of red but of the grammar of the word "red." Similar considerations apply to the statements, "It is a property of the number 1 that it is had by a lecturer in this room" and "It is a property of 1 that $1 < 2$". In *Principia* Russell talks of individuals and properties after the model of ordinary language.

4 The definition of class equality by means of 1-1 correlation raises the question whether the classes must in fact be correlated with the paradigm in order to have the same number, or whether this need only be possible. What is the criterion for there being a *possibility* of correlating them? Is it that if you tried you would succeed? If so, then that there are, for example, three crosses provided you *can* correlate them with the paradigm, leaves the crosses *hypothetically* correlated. Hence the criterion of the possibility of correlation is their actually being correlated.

There is something queer about Dedekind's definition of an infinite class: a class is infinite if it can be correlated 1-1 with a proper subclass of itself, finite if it cannot be. Giving a criterion presupposes that we can use it, and this implies that we can *try* to use it. But what does it mean to try to correlate a proper subclass to the class? Dedekind has not given a criterion which one could *use* to distinguish finite from infinite. It seems nonsense to say that an infinite class is such that it could be 1-1 correlated to a proper subclass. We have not tried nor could we try, to correlate in this sense. *It makes no sense to try.* We mean something utterly different by "correlation" when we speak of correlating the infinite class of cardinals with the odd numbers. "Correlation" is being used in a new sense. The point may be put in this way: the correlation of an infinite class with a part of itself, e.g., 1,2,3,4.... with 1,3,5,7..., is a different correlation when the words "and so on" are added. We correlate 1,2,3 with 1,3,5 in the old way and give a law besides. Similarity and equality both mean something different here, though there are analogies. In fact, although the symbol 1,2,3,4... is entirely different from 1,2,3,4, there are analogies, as is evidenced by the way of writing it. Some of the rules are common to finite and infinite sets. And of course some are different: 1,2,3... = 1,2,3,4...

If we have a law holding for all cardinal numbers, and hence one which cannot be tested by going through the entire series, our inability to carry out the test is sometimes said to be due to human weakness. I want to say that it has no meaning at all to assert that we are too weak to go through the cardinals. In asserting this, what we are doing is comparing writing down all the cardinals with writing down 1,000,000 of them on a small blank card. The two are impossible in two entirely different senses, the former because we cannot perform something corresponding to nonsense. Now a series can be said to be of infinite length if there is a method of measurement. The sense of this statement, like that of "This rod is three yards long," depends on how we determine its length, and differs according to the method of measurement. A method of measurement must be given before the statement can make sense. Propositions which cannot be verified are not necessarily useless, e.g., that a comet describes a parabola for 10,000 years. But it makes no sense to say that because people do not live long enough there is no final test for "The comet describes a parabola." No means of verifying the whole path has been provided. Note how different is the statement "It describes a parabola for three years" from the statement "It describes a parabola."

5 When we construct an arithmetic we have the idea that it will include propositions involving the general concept of *number* and the concepts of *odd number* and *even number,* and that an arithmetic making no mention of these would be incomplete. In particular, that if we had a calculus with the numerals and multiplication there would have

to be a *general* law in order to proceed to a new case, and that the general law would obviously contain the term "cardinal number." But an explanation of how to calculate does not require a general expression. Note that in teaching children arithmetic we need not mention the general concept *number*. We teach with *particular* numbers, and in our explanation of how to calculate we need not have any general expression. A calculus without general expressions is not less complete; it is just another game. (One could write a book of games in which the word "game" did not occur.) Chess is complete without any added complications. Added complications make a fresh game.

6 What is the meaning of sentences of the form "There is a number satisfying so-and-so" or "There is a number with such-and-such a property?" And in what sense is a proof by induction a proof of a general proposition, e.g., "For all cardinal numbers such-and-such is the case?" To consider questions of this sort it is not necessary to begin with sentences using the notation $(\exists x)fx$ and $(x)fx$. One notation is not more exact than another. A notation may be more elegant but not more exact. Knowing a notation is exact is knowing what it can do. In considering the question, What does the proof that $1{:}3 = 0.333\ldots$ prove?, we shall begin with the common notation for division of 1 by 3. One answer to this question is that it proves that there will be an infinite number of 3's or no other numbers than 3. The proof is that remainder = dividend. The recurrence of 1 in the division

$$\begin{array}{r} 3)1.0\ (.3 \\ \underline{9\quad\ \ } \\ 1\quad\ \ \end{array}$$

shows that the 3's go on. Suppose this recurrence had not been noticed, and someone asked, "Are there always to be 3's, or will another number appear sometime?" Without noticing the recurrence one would have no way of answering this question. Now in what sense can discovery of a recurrent dividend be an answer to the question whether there will always be 3's, which was concerned with an infinite extension? The answer seems to say nothing about an infinite extension. In what way will knowing the proof $\frac{1:3}{\underline{1}}$ put an end to the investigation? *Must* it end? Of course one could go on dividing and get a 4 by mistake. Why is it that we can prophesy that there will never be a 4, that any 4 we get would be a mistake? And does the statement, "There will never be a 4" have a sense apart from the proof, or does the proof give it a sense?

Our present example resembles our getting the notion of *odd number* and *even number* on being taught specific numbers. By having our attention directed to something not noticed before we get hold of something new. Suppose we had been taught division without the notion of periodicity. We would have a complete calculus without this notion; but writing the symbol $\dot{3}$ for the law that remainder = dividend when we notice the recurrence of 1's, introduces something entirely new. We actually have in $\dot{3}$ a new symbol.* $\frac{1:3}{\underline{1}}$ is a new operation and has a result in a different sense than $\frac{1}{3}$. 0.333... seems to refer to an extension (.3, .33, .333,...), while 0.3 does not. The latter is not an extension, [nor is it an abbreviation].†

Before there was a proof of the irrationality of π, the question whether there is a recurrence in the expansion of π was not clear until one had a method for determining it. In the case of people who had no method at all, e.g., people who were even surprised that 1:3 continued to repeat, the question whether $\frac{1}{3}$ is periodic and the question whether π is periodic are alike. If one thinks of these questions as referring to an extension, to scores of decimals, one would think the method of continued calculation a good one. But it would not be a method for determining whether there would *always* be recurrence. The question whether there will always be recurrence derives its sense from the answer to it. "Is 1:3 recurrent?" has an answer; but "Is the expansion of π recurrent?" has none so long as one's method is to look for a recurrence in the infinite development. This method mixes up looking for a recurrence with looking for one in a finite interval.

If it does not strike us as queer that proving 1:3 to be recurrent by giving the law that the remainder equals the dividend as an answer to the question, "Will the 3's always continue?," which appears to concern an infinite extension, the reason is this: that 1:3 is taken to be a symptom of the recurrence of 3's in the infinite extension, as jaundice is taken to be a symptom of trouble in the liver. Are these analogous? In what way can we say $\frac{1:3}{1}$ is a symptom of continuing recurrence? What would you say to a person who did not see immediately that *it*

* See *Philosophische Grammatik*, p. 404. (Editor)

† *Ibid.*, p. 428. (Editor)

was a symptom? The explanation would be: 1 divided by 3 leaves 1, which being the same as the beginning number will when divided by 3 leave 1, *and so on*. The italicized phrase is the second part of the explanation showing the connection between $\frac{1:3}{1}$ and recurrence.* But this explanation does not supply the extension.†

If we watch a man dividing 1 by 3, then the question whether he will always write 3's is like a question of physics—like asking whether a comet will always describe a parabola. The mathematical question whether 3 recurs, in contrast to the question about what the right-hand side of the equation $1:3 = 0.33$ .. will look like as he continues to divide, is a question about the *whole equation*. To say that 1:3 *yields* a recurrent 3 is not to say something about the fate of 0.33.. but about the calculation $\left(\frac{1:3}{\underline{1}}\right)$ by which what is meant by "recurrence" is defined. The result of a mathematical proof gets its meaning from the proof.

It is useful to compare the proof that 1:3 is periodic with the proof that $\sqrt{2}$ is irrational. The question about recurrence in the cases of 1:3 and 1:5 is the same, but in the case of $\sqrt{2}$ it is very different. Given that $\sqrt{2} = 1.414\ldots$, if someone asks the question "Is 14 recurrent?", you can ask in turn, "In what cases would you say it recurs?" Your question will get from him what his question meant. If he replies, "I shall say it doesn't recur if the next figure is not 1", then you get what he means by his question. He should tell you how it is to be answered. If the question is whether there will be *any* periodicity, there would be no sense to it because it has not been given any sense. It is like asking about the length of something when measured by a rod that changes from 0 to an indefinite length—though this question does not even have a false answer as does the question whether 14 recurs. Now in the case of 0.33... you have given criteria for truth and falsity of the statement that 3 recurs: (a) for truth, $\left(\frac{1:3}{1}\right)$ (though this is not really a criterion for the answer to the question asked, "Will there always be 3's?"); (b) for falsity, the appearance of some other number, say 4. But if 4 occurred, you would not be satisfied and would question

* See *Philosophische Grammatik*, pp. 283–4. (Editor)
†*Ibid.*, pp. 427–29. (Editor)

whether the calculation had been done correctly. 4 must come by a *correct* calculation. And to decide whether it is correct one must look at the construction $\left(\begin{matrix}1:3\\1\end{matrix}\right)$ and not at 0.33...

If we fix the proof of the proposition that there will always be 3's, we must have come to a decision between two possibilities: our result, "There will always be", and its negative "There will not be". The answer to the one is that the first remainder = the dividend, and to the other that it does not. We now cannot find an analogy between $\sqrt{2}$ and 1:3, for the calculations in the two cases are entirely different. The only analogy is that the right-hand sides of the equations are somewhat alike. If to the statement "1:3 yields a recurrent 3" one has given a meaning, then prior to the proof of the irrationality of $\sqrt{2}$ one has given no meaning to the statement that $\sqrt{2}$ does, or does not, yield a recurrent 14. To define the meaning of a statement we must define its negative. We tend to think that here we know what we mean by nonrecurrence because we have a finite series in mind: $\begin{matrix}3434343\\3435678\end{matrix}$. In the case of $\frac{1}{3}$ we have only a few figures, but recurrence is defined by $\begin{matrix}1:3\\1\end{matrix}$. In the case of $\sqrt{2} = 1.414..$, however, we have drawn no such distinction between recurrence and nonrecurrence as we have by the construction $\left(\begin{matrix}1:3\\1\end{matrix}\right)$.

7 A person who is taught to multiply can ascertain that $16 \times 16 = 256$. Does he know that $256 \div 16 = 16$? He does not, unless $\frac{256}{16}$ means just $16 \times 16 = 256$. Suppose one is taught the multiplication of two recurrent decimals, e.g., $.424242.. \times .3636...$ The answer will be the result of the multiplications 0.42×0.36, 0.4242×0.36, etc. The middle part of the periodic decimal does not change in the course of these multiplications, although there is no periodicity in the end figures. Having been taught multiplication of recurrent decimals, suppose the problem is reversed and one is asked whether division of 1 by 3 gives a recurrent decimal. The right-hand side of the equation $1:3 = ...$ has not been given a sense, inasmuch as "recurrent decimal" has a meaning only in the calculation $0.4242... \times 0.3636...$, where we multiplied these decimals. The latter multiplication of decimals will not help at all.

The question has been raised whether $0.333.. \times 3.0$ would not be a proof that 1:3 is recurrent. (Assume that in being taught multiplication of recurrent decimals you were taught that $0.333... \times 3 = 1.0$ and not 0.999...) If this is the sort of thing you would accept as an answer, it *is* an answer. But this does not mean that a person can divide 1 by 3. We have here two different calculi and hence two results. The point of the question raised here is that if one could arrive at the result of periodic *division* by multiplication, this contradicts my statement that the result of periodic division is bound up with the process. We seemingly have got the same result here in different ways. That is, the question, "Is 1:3 periodic?", is seemingly answered in two ways. But I deny that it is the answer to the *same* question. Consider an analogue: $14 \times 15 = 210$ and $210:14 = 15$. The question to which division is an answer is a different question. By teaching that the first can be written as the second, we have not taught division nor have we taught the answer resulting from division. "$14 \times 15 = 210$" gets its sense from the calculation. Demonstrations which are said to prove the same thing usually meet only in the result and have no rapport before that point.

If 1:3 is all that has been taught, the question, "Does a:b give a periodic decimal?", has no sense except where $a = 1$ and $b = 3$. Suppose periodic division of 1 by 3 had been taught, then the following four questions concerning it can all be answered, as there is a criterion for each answer: (1) Is there a recurrent period?, (2) What is it?, (3), Is it *this* period?, (4) Is it a disorderly decimal? Suppose one was given merely that $1:7 = 0.14$. You cannot ask question (1), as you have no calculus, nor can you ask whether it is a disorderly decimal. We tend to suppose the same questions can be asked sensibly in all cases. Here we have a case where the calculus has not given sense to the answer. You can't ask "Is there a period?", but only "Will 14 recur?" This is not due to a gap in your knowledge but to a gap in your calculus. Whoever gets a method for looking for a period learns a new calculus. In general one cannot ask whether 1:b gives a disorderly development unless this is merely another way of asking whether it is periodic. The fact that $1:3 = 0.3$ answers all four questions shows the sense of the questions. *The sense is determined by the method of solving.* To the question there corresponds a general law for finding answers.

If you want to explain what you mean by measurement, tell me your method of going about it. The construction of $\sqrt{2}$ by the following

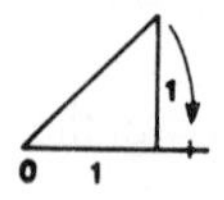

is a way of measuring a length on the base line. People have said that they had found a point on the straight line which is not a rational distance from 0. The idea is that $\sqrt{2}$ is the result of the construction, i.e., is a certain length, whereas it *is* the construction. It is absurd to say that $\sqrt{2}$ is the length on the base, for the length is what it measures. Hence accuracy does not come in, for that has to do with measuring rods. Nor is it an approximation. A construction such as this is a calculation, a symbolism.

8 Hardy begins his discussion of the general definition of a real number with an account of the special case of defining $\sqrt{2}$, as follows:

> The square of any rational number is either less than or greater than 2. We can therefore divide the positive rational numbers . . . into two classes, one containing the numbers whose squares are less than 2, and the other whose squares are greater than 2. We shall call these two classes *the class L,* or *the lower class,* or *the left-hand class,* and *the class R,* or *the upper class,* or *the right-hand class* . . . Every member of R is greater than every member of L, we can find a member of L and a member of R whose square differs from 2 by as little as we please, and L has no greatest member and R no least member . . . This mode of division of the positive rational numbers x into two classes, such that $x^2 < 2$ for the members of one class and $x^2 > 2$ for those of the other . . . is called a section . . . We denote the section or number thus defined by the symbol $\sqrt{2}$ . . . A section of rational numbers, in which both classes exist and the lower class has no greatest member, is called a real number.*

Concerning the division of the infinity of rationals into two classes, given a principle of division we can say the square (or cube, or fourth power) of certain numbers is smaller than a given number. But has the word "section" any meaning that passes beyond specific powers and specific numbers, e.g., $x^3 < 2$? Are Hardy's examples only for beginners, so that we could have the general calculus without these? Are the examples *essential?* [His general discussion of sections enumerates three mutually exclusive possibilities, that R has a least r, L a greatest l, neither section has a least or greatest member.] These general terms, R, r, L. l, get their meaning only from such examples as are given, viz., $x^2 < 2$, $x^2 > 2$. All that has been explained to us as a section is the numbers satisfying these functions. Also, what sense is

*G. H. Hardy, *Pure Mathematics,* Cambridge University Press, 5th ed. 1928, pp. 7–19. This, or an excerpt of comparable detail, was read out in the lecture. (Editor)

there to the symbol "*P*", which denotes a property belonging to all rational numbers, if no examples are given? Hardy gives one property, $x^2 > 2$, and then talks of "other properties", to which his calculus has given no meaning, for example, "being rational". What is the property of being rational—rational as opposed to what? Perhaps to cardinal.

Mathematically, Hardy has done one of two things, either (1) he has given "*P*" the meaning which the example gives it, in which case we have only a less intelligible language, or (2) we have a brand new calculus with these general terms *P, Q, L, R*. To begin as though the examples were for the stupid and then to talk of proceeding generally, dividing all rational numbers into the classes *L* and *R*, has no sense. The general terms *L* and *R* do not extend the field one starts with; they are a new type of term. We have a new calculus with these general terms, and the new calculus does not represent the discovery of a larger field. We have a new field. If examples are not essential we can replace the words "property", "section", "upper", "lower" by new words since we become muddled when we use familiar words. Then we would have one calculus with the new words and another with $x^2 > 2$, $x^2 < 2$.

When Hardy speaks of the three mutually exclusive possibilities, one possibility being that neither section has a least or greatest member, this leaves the impression that a definition of "real number" could be given by general considerations, with examples only for beginners. But if we did not have the example $\sqrt{2}$ to explain one class having no greatest member and the other no smallest, we could not define "real number". Before introducing $\sqrt{2}$, talk of such a class in connection with cardinals would be nonsense.

A general approach can make it appear that the particular cases are examples of an idea, whereas the examples are rigorously needed to explain the idea. In the differential calculus examples would be needed if one defined a function $f(x)$ by teaching what sine x, cosine x, etc. are, then saying that $f(x)$ is one of these and perhaps others. $f(x)$ is then a shorthand for these. If examples are not needed, then $f(x)$ can be explained independently of any particular function, and must have meaning to a person who had never heard of cosine x, etc.

Let us look at Hardy's statement that since any rational number is a mode of division of the rational numbers we can substitute for it the section. For example, ½ could be replaced by the section (pair of classes) produced by ½. Now what does it mean to replace the number by a section? For purposes of comparison consider only the cardinals 0,1,2,3,4..., and the section we shall call 3, viz., the section following it (or including and following it). Suppose instead of 3 we introduce a new sort of number for the section: the number together with two others plus dots. 3 = 3,4,5... Calculation of 3 × 6 would be represented: 3,4,5... × 6,7,8... = 18, 19, 20... These new numbers would then have the same calculus as the old. The numbers following 3, 6, and 18 might just be considered otiose [so that there has really been no replacement of 3, 6, etc. by new numbers; we simply calculate with 3 and 6.] If Hardy says he could calculate with the classes just as well as with the rationals, there has really been no substitution at all. Calculation is simply with the rationals. But when he talks of sections to which no rational number corresponds, we only have a generalization of our system insofar as the notions of greater and smaller are similar to these notions in our systems of rationals.

Consider the statement "A rational number is comparable to all other rationsls". This means there is a way of finding out. Now examine "Consider *anything* that is comparable to a rational number". This means nothing, for we have defined "comparable" only for the rationals, e.g., for 4 and 3 and the fractions ¾ and ½. To say $\sqrt{2}$ is smaller or greater than some rational requires a new definition of "comparable". $x^3 < 2 = x < \sqrt[3]{2}$ is a new definition of "smaller". The old notion of "smaller" comes in, but something different as well.

Hardy says of "$\frac{1}{2} < \frac{1}{3}$" that it is ambiguous in that it may be interpreted as a statement of elementary arithmetic or as a statement about sections of rationals. I want to say that what interprets the formula "$a > b$" is the calculus surrounding it. To give a new interpretation of "a" and "b" is to place them in new surroundings. And to see whether in fact we are dealing with a new interpretation, look at the calculus. If the calculus is the same, then we have no new interpretation. (When we let 3 = 3,4,5..., inasmuch as the calculations with the two are the same, 3 and 3,4,5... do not differ.) Hardy said that the context in which "$\frac{1}{2}$" occurs may be sufficient to fix the interpretation, that "$\frac{1}{2}$" in "$\frac{1}{2} < \sqrt{\frac{1}{3}}$" must be the real number $\frac{1}{2}$. I want to say that what has changed is the meaning of the sign "<". My general definition of a new sign "$\leqslant$" shows its relation to "<": $x \leqslant \sqrt{\frac{1}{a}} \overset{Df.}{=} x^2 < \frac{1}{a}$. In the present example, $\frac{1}{2} \leqslant \sqrt{\frac{1}{3}} = \left(\frac{1}{2}\right)^2 < \frac{1}{3}$.

Suppose we wrote all rationals as infinite (recurrent) decimals, that is, replaced a rational by a process of approximation. $\frac{1}{2} = .4999\ldots$ Have we replaced it by a different *entity?* No. We have merely another notation. It has been asserted that the two lines [0 1/4 1/3 1/2 1], [0 ¼ ⅓ ½ 1 $\sqrt{2}$], the one with rational numbers and the other with real numbers, are composed of different entities, the second not being a mixture of the old and something new. The entities between 0 and 1 on the two lines are said to correspond but not to be the same. Similarly, that among decimals some are terminating, some periodic, and the latter are of two sorts, e.g., $.4\dot{9}$ and $.\dot{3}$. In reply, I want to say that .5 and the approximation $.4\dot{9}$ are not different things. "$.4\dot{9}$" is just a different notation. In *general* proofs where we employ .5 and $.4\dot{9}$, a new general notation is required. $\frac{1}{2} = .5 = .4\dot{9}$ *alludes* to a different idea, but we do not have here different things. *Wherever* in a calculus one number can be replaced by another, as 2 by $2 + (0 \times i)$, they are the same. A proof mentioning "any decimal", while not a proof about $.4\dot{9}$ and .5, could be applied to both equally. A proof *with* .5 must be the same as the proof with $.4\dot{9}$, since whatever approximation corresponds to $\frac{1}{2}$ must have the arithmetic properties of $\frac{1}{2}$. But a proof *about* $.4\dot{9}$ may be different from one about .5 because the two proofs allude to a different general notation.

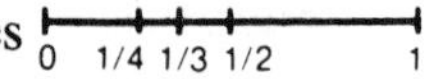

Hardy applies to the aggregate of *real* numbers the method applied to the rationals. [For sections of real numbers the "mode of division" must now be explained, and the question again arises whether examples are essential to the general calculus.] The mode of division can be explained either (1) by taking an example, such as $\sqrt{2}$, as the sort of thing meant by a mode of division, or (2) by the explanation's being contained in the general theory. Hardy explains "mode of division of the real numbers" by the example of a real number, $\sqrt{2}$. P and Q are mutually exclusive properties possessed by any real number. P might be $x \leq 2$ and Q, $x > 2$. Here L has a greatest member l or R a least member r, but both of these possibilities cannot occur, as they could with the sections of rationals. We have only the calculation $\frac{l+r}{2}$ to show this*.

I am startled at the ease with which continuity is investigated. With practically no reasoning or examples we arrive at the continuum.

In connection with the *a priori* enumeration of possibilities in the course of defining a real number, namely that the L class has a greatest member or the R class a least or that neither class has a greatest or a least, it makes no sense to point out that certain possibilities are realized. For we do not have here a case of possibilities which are realized in particular instances. Only when $\sqrt{2}$ is introduced do we give sense to the possibility of one class having no first member and the other no last. That any rational number can be called a principle of division of the rationals is clear. But division is a very different matter when L has no greatest member and R no least. [When a real number is a principle of division among classes], the *a priori* enumeration of possibilities, with one possibility ruled out, has no meaning. The word "division" introduces a simile. Division is not like dividing a cake. How far can it be used?

When Hardy says he will apply the same method of division to the aggregate of real numbers as to the rationals, [thereby extending the the concept of division of rational numbers to real numbers], the picture used is

where | | are rationals, | is division, and • • • are the real numbers which fill the gaps between

[* If L had a greatest member l *and* R a least member r, then $\frac{l+r}{2}$ would be greater than all members of L and less than all members of R and so could not belong to either class. Any section of real numbers, according to Hardy, will correspond to a real number in the same sense that a section of rationals sometimes, though not always, corresponds to a rational number. An important difference is that the idea of a section of rationals led to a new idea of number, that of a real number, more general than that of a rational number, whereas the idea of a section of real numbers does not lead to a more general conception of number. . . . The aggregate of real numbers is called the arithmetical continuum.]—These remarks are from Hardy's account of real numbers in *Pure Mathematics*, in large part in his words, and are included so as to provide a context for Wittgenstein's comments. (Editor) the rationals. ["Our common-sense notion of the attributes of a straight line, the requirements of our elementary geometry and our elementary algebra, alike demand the existence of a number x greater than all the members of L and less than all the members of R, and of a corresponding point P on the line such that P divides the points which correspond to members of L from those which correspond to members of R."* One gets the idea that otherwise there must be a point left out on the straight line.†] If people had always painted geometrical figures with a brush, so that the dividing line between colors was a line and an intersection a point, they would never have got the notion of a class of points. The figures

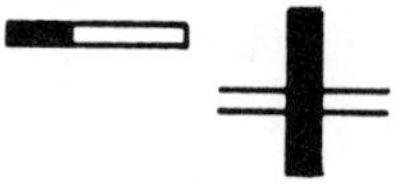

would not have suggested a point belonging to one or the other classes, whereas dots do. The simile of dividing classes of points is connected with our peculiar way of drawing points and lines. Unless we had the idea given by the first picture above we would not say it is possible to apply the same method of division again to the real numbers as Hardy suggests.

9 Suppose we divide a line AD in accordance with the rule of bisecting the interval on the left if we throw tails and on the right if we throw heads. For example:

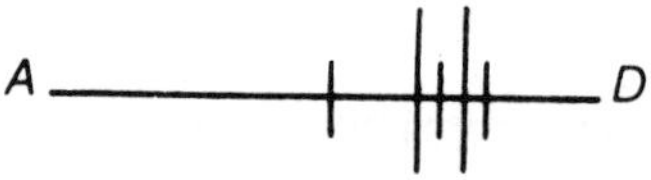

A D

Here we believe ourselves to be determining a point by ever decreasing intervals in which it lives by repeatedly throwing the coin and thereby always diminishing the abode of the point. Also, we take it that the point corresponds to an irregular infinite decimal. But by throwing a coin so as to decrease the interval, we have not determined a point endlessly approached by the cuts made in accordance with the repeated tosses of the coin. We have really a series of intervals, which will always remain such. After every throw the point is still infinitely indeterminant.‡ The trouble is with our imagery.

*G. H. Hardy, *Pure Mathematics*, p. 9. (Editor)

†See *Remarks on the Foundations of Mathematics*, Oxford, and Cambridge, Mass., 1967, p. 151. (Editor)

‡*Philosophische Grammatik*, p. 477. (Editor)

We should make a distinction between a class of tosses, or mere choices, and a way, or rule, for making choices. The latter defines an irrational number. An irrational number is a process, not a result. We have a tendency to think that there is one result produced by $\sqrt{2}$, viz., an infinite decimal fraction. $\sqrt{2}$ produces a series of results, but no single result. $\sqrt{2}$ is a *rule* for producing a fraction, not an extension. Now there is this difference between the rule for constructing a decimal by repeated throws of a coin and the rule for working out places of $\sqrt{2}$, namely, that we have a fixed method for deciding for any rational number whether it is larger or smaller than $\sqrt{2}$. $\sqrt{2}$, i.e., the rule, *is* a point, but only because we have this method by which we can calculate as with the rational numbers.*

10 Suppose a binary fraction is constructed by writing 1 whenever $x^n + y^n = z^n$ and writing 0 otherwise, letting x, y, and z be within the range 0—100. And suppose someone asked whether .11000... is larger than or equal to .11. There is no answer, for we have arranged no method of answering. The same situation obtains for the problem of finding the distribution of primes. If a question is asked for which there does not exist a method of solution, does the question have meaning? I have said No, and have likened the conclusion of a mathematical proof to the end surface of a cylinder. The proved proposition is the end surface of the proof, *a part of it*. Similarly, the result of a construction is not something by itself; the construction is essential to it. For example, the construction of $\sqrt{2}$ as a measure of a length on the base line:

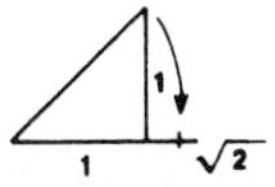

Without the construction $\sqrt{2}$ is not the length. This length is not an approximation. It has nothing to do with measurement by a foot rule.

In what sense can one say that a question in mathematics makes no sense? It would seem that if it does not make sense, we could never know where the answer lay. Ask yourself, What uses does one make of the question? It does stand for a certain activity by the mathematician, of trying, of messing about. If the question did not stand for something, one would expect *any sort* of activity. The question has then that meaning—as much meaning as the messing about has. The

*See *Philosophische Grammatik*, pp. 484–5, for fuller discussion of the comparison. (Editor)

mathematician's activity is carried on in a particular sphere. A question is part of a calculus. What does it prompt you to do? When a question is asked for which there is no method of answering it, we do know certain requirements the answer must fulfill. In one sense it is true to say this, but it is misleading. When Hardy says he believes Goldbach's theorem, I would ask him what his belief in this theorem led him to. What does he do? It may have led him to attempts to prove it, which shows that *some* meaning attaches to the theorem inasmuch as these activities would not have been caused by another theorem.

Suppose that two people who were set the same problem got different solutions. Then for one person to show the other that the two proofs ought to come to the same result it would need to be shown how the calculi meet, i.e., one calculus would need to be made. What does it mean for two proofs to prove the same verbal form? You may find on looking *closely* that they prove similar things, but not the same thing. How is it to be decided whether they prove the same thing? You must look at the proofs to decide.

What is the system in which we say that every equation has a root? Is there a system where this might not be the case? The proof that *there are n* roots of an equation of degree *n,* that they exist even though we have no method of finding them, is queer in the way a proof of the construction of a pentagon would be queer if it did not tell us how to construct a pentagon. The phrase "proof of existence" has a different sense here than it does where there is such a method. We say we "think we know what we mean by 'root of an equation of 10th degree' ". But do we? "Root" has meaning in terms of a proof in which it functions. [". . . the proposition 'This equation has *n* roots' hasn't *the same meaning* if I've proved it by enumerating the constructed roots as if I've proved it in a different way. If I find a formula for the roots of an equation, I've constructed a new calculus . . ."*]

It has been observed that the expression of the result of a proof is a way of cataloguing the proof. The concluding *prose* sentence is in a particular sense a short-hand of the proof. It would be like a title, which has a definite relation to the text of a book. If a proof deserves the title, i.e., the end result, then the result stands for the proof. When it does, the final sentence of a demonstration is not like a proper name, else it would be the name of the demonstration. It is part, namely the end, of the proof, and the proof incorporates it into a new calculus.

**Philosophische Grammatik,* p. 373. (Editor)

Were the proof to stand in the same relation to the conclusion as verifications do to the statement that Smith is in his room, then we could call the proof a sort of symptom of the conclusion. But in mathematics the proof is not a symptom, for the proved proposition is part of the proof. The concluding prose sentence can serve to catalogue the proof by being part of a system of language. If you want to know the function of what is called the result of a proof, see how far it catalogues the proof, or whether it is only a name. ["The verbal expression of the allegedly proved proposition is in most cases misleading, because it conceals the real purport of the proof, which can be seen with full clarity in the proof itself."*]

11 Let us look at considerations which led to Russell's theory of types.

Let $f(a) = U$'s coat is red
$F(a) = U$'s coat is a color of the rainbow
$\phi(f) =$ Red is a color of the rainbow.

Now does $\phi(F)$ have meaning? Russell would say that "a color of the rainbow has the property of being a color of the rainbow" does not have meaning, and in general that "$f(f)$" does not. Now if we make a rule of grammar excluding one substitution, which is what the theory of types does so as to avoid contradiction, we must not make it depend on a property of anything but symbols. We must give a formal criterion forbidding it: that when we introduce "$f(x)$" we are not thereby allowed meaning to "$f(f)$". Consider $\sim f(f) = F(f)$, and the expression got by replacing "f" by "F": the property of not having itself as a property has itself as a property. From $\sim f(f) = F(f)$ the contradiction $F(F) = \sim F(F)$ results. The root of the contradiction is in making a function a function of itself. That the result is a contradiction means that "f" cannot be used as an argument in "$f(x)$".† But why should it not come out this way, inasmuch as what you started with is no proposition?† It is not right to say the law of contradiction has been violated, for this could only be the case if you were talking of propositions. We merely have a game here that leads to something that looks

**Philosophische Grammatik*, p. 370. (Editor)

† See *Remarks on the Foundations of Mathematics*, p. 178, and *Zettel*, Ludwig Wittgenstein, Schriften 5, Suhrkamp, Frankfurt am Main, 1970, p. 424.

like a contradiction. You can either say "$f(f)$" is meaningless, or that "f" outside the bracket stands for a function of higher order.

Hardy said it would be intolerable to have real numbers of different orders. See his discussion of the upper bound of a sequence of real numbers as being of a different order because it is defined by reference to a totality of which it is the bound. An analogous example is the maximum of a curve, defined as the highest of all points on the curve. The axiom of reducibility says that a number of higher order can be calculated by processes which define numbers of lower order. This axiom is like a proposition of physics; it seems to be *true*. Now we cannot have an axiom which will have to be, or which can be, borne out by a special case. A mathematical axiom about number is a postulate or rule according to which we proceed. The axiom of reducibility states that there is a number of lower order even though there is no way of calculating it, e.g., that an irrational that is a maximum of a curve exists though there is no rule for constructing it. I ask, What is the character of a real number? If it is a method of developing a decimal fraction infinitely, there is no room for the idea of different orders. Processes of developing a decimal fraction are of the first order. Thus π and e might be considered to be numbers of the first order, and π^e as a process of second order. (π as a series of rationals, and π^e as a series of irrationals). A series of irrationals is an irrational of the second order, as a series of rationals is an irrational of the first order. Now we have a process of developing π^e just as for developing π. And it is in no way objectionable to have what one might call a process of the second order, though one might as well call it a process of the first order since one can write down the development of it. We have a process of the first order wherever we can write down a development in the decimal system. It is as though the difficulty comes in *before* the process of development. The axiom of reducibility says that there is a development, say, of an irrational, e.g., the maximum of a curve, though no process of development has as yet been found. A number that we have no method of developing is a number in a different sense. In the case of an irrational number without a development we supposedly have a description corresponding to which there is a number which can be found by looking for a method of development; and this number will be the irrational number described. Discovery of this number is treated analogously to making an expedition of discovery or solving a problem in physical science by finding something corresponding to a description. But the analogy is misleading.

What counts in mathematics is what is written down. Symbols obviously do interest even the intuitionist, who says that mathematics is not a science about symbols but about meanings—just as a zoologist might say, analogously, that zoology is not a science about the word "lion" but about lions. But there is no analogy between mathematics and zoology in this respect. The intuitionist should be asked to show how "meaning" operates. In Chapter II of *Grundlagen der Arithmetik* Frege attacks formalism. But there is this much correct about formalism: if a mathematician exhibits a piece of reasoning one does not inquire about a psychological process.

There is no retreat in mathematics except in the gaseous part. (You may find that some of mathematics is uninteresting—that Cantor's paradise is not a paradise.)

The talk of mathematicians becomes absurd when they leave mathematics, for example, Hardy's description of mathematics as not being a creation of our minds. He conceived philosophy as a decoration, an atmosphere, around the hard realities of mathematics and science. These disciplines, on the one hand, and philosophy on the other, are thought of as being like the necessities and decoration of a room. Hardy is thinking of philosophical opinions. I conceive of philosophy as an activity of clearing up thought.

GREAT BOOKS IN PHILOSOPHY PAPERBACK SERIES

ESTHETICS

- ❑ Aristotle—*The Poetics*
- ❑ Aristotle—*Treatise on Rhetoric*

ETHICS

- ❑ Aristotle—*The Nicomachean Ethics*
- ❑ Marcus Aurelius—*Meditations*
- ❑ Jeremy Bentham—*The Principles of Morals and Legislation*
- ❑ John Dewey—*The Moral Writings of John Dewey, Revised Edition* (edited by James Gouinlock)
- ❑ Epictetus—*Enchiridion*
- ❑ Immanuel Kant—*Fundamental Principles of the Metaphysic of Morals*
- ❑ John Stuart Mill—*Utilitarianism*
- ❑ George Edward Moore—*Principia Ethica*
- ❑ Friedrich Nietzsche—*Beyond Good and Evil*
- ❑ Plato—*Protagoras*, *Philebus*, and *Gorgias*
- ❑ Bertrand Russell—*Bertrand Russell On Ethics, Sex, and Marriage* (edited by Al Seckel)
- ❑ Arthur Schopenhauer—*The Wisdom of Life* and *Counsels and Maxims*
- ❑ Benedict de Spinoza—*Ethics* and *The Improvement of the Understanding*

METAPHYSICS/EPISTEMOLOGY

- ❑ Aristotle—*De Anima*
- ❑ Aristotle—*The Metaphysics*
- ❑ Francis Bacon—*Essays*
- ❑ George Berkeley—*Three Dialogues Between Hylas and Philonous*
- ❑ W. K. Clifford—*The Ethics of Belief and Other Essays* (introduction by Timothy J. Madigan)
- ❑ René Descartes—*Discourse on Method* and *The Meditations*
- ❑ John Dewey—*How We Think*
- ❑ John Dewey—*The Influence of Darwin on Philosophy and Other Essays*
- ❑ Epicurus—*The Essential Epicurus: Letters, Principal Doctrines, Vatican Sayings, and Fragments* (translated, and with an introduction, by Eugene O'Connor)
- ❑ Sidney Hook—*The Quest for Being*
- ❑ David Hume—*An Enquiry Concerning Human Understanding*
- ❑ David Hume—*Treatise of Human Nature*
- ❑ William James—*The Meaning of Truth*
- ❑ William James—*Pragmatism*
- ❑ Immanuel Kant—*The Critique of Judgment*
- ❑ Immanuel Kant—*Critique of Practical Reason*
- ❑ Immanuel Kant—*Critique of Pure Reason*
- ❑ Gottfried Wilhelm Leibniz—*Discourse on Metaphysics* and the *Monadology*
- ❑ John Locke—*An Essay Concerning Human Understanding*
- ❑ Charles S. Peirce—*The Essential Writings* (edited by Edward C. Moore, preface by Richard Robin)
- ❑ Plato—*The Euthyphro*, *Apology*, *Crito*, and *Phaedo*
- ❑ Plato—*Lysis*, *Phaedrus*, and *Symposium*
- ❑ Bertrand Russell—*The Problems of Philosophy*
- ❑ George Santayana—*The Life of Reason*
- ❑ Sextus Empiricus—*Outlines of Pyrrhonism*
- ❑ Ludwig Wittgenstein—*Wittgenstein's Lectures: Cambridge, 1932–1935* (edited by Alice Ambrose)

❑ John Maynard Keynes—*A Tract on Monetary Reform*
❑ Thomas R. Malthus—*An Essay on the Principle of Population*
❑ Alfred Marshall—*Principles of Economics*
❑ Karl Marx—*Theories of Surplus Value*
❑ David Ricardo—*Principles of Political Economy and Taxation*
❑ Adam Smith—*Wealth of Nations*
❑ Thorstein Veblen—*Theory of the Leisure Class*

HISTORY

❑ Edward Gibbon—*On Christianity*
❑ Alexander Hamilton, John Jay, and James Madison—*The Federalist*
❑ Herodotus—*The History*
❑ Thucydides—*History of the Peloponnesian War*
❑ Andrew D. White—*A History of the Warfare of Science with Theology in Christendom*

LAW

❑ John Austin—*The Province of Jurisprudence Determined*

PSYCHOLOGY

❑ Sigmund Freud—*Totem and Taboo*

RELIGION

❑ Thomas Henry Huxley—*Agnosticism and Christianity and Other Essays*
❑ Ernest Renan—*The Life of Jesus*
❑ Upton Sinclair—*The Profits of Religion*
❑ Elizabeth Cady Stanton—*The Woman's Bible*
❑ Voltaire—*A Treatise on Toleration and Other Essays*

SCIENCE

❑ Nicolaus Copernicus—*On the Revolutions of Heavenly Spheres*
❑ Charles Darwin—*The Autobiography of Charles Darwin*
❑ Charles Darwin—*The Descent of Man*
❑ Charles Darwin—*The Origin of Species*
❑ Charles Darwin—*The Voyage of the* Beagle
❑ Albert Einstein—*Relativity*
❑ Michael Faraday—*The Forces of Matter*
❑ Galileo Galilei—*Dialogues Concerning Two New Sciences*
❑ Ernst Haeckel—*The Riddle of the Universe*
❑ William Harvey—*On the Motion of the Heart and Blood in Animals*
❑ Werner Heisenberg—*Physics and Philosophy: The Revolution in Modern Science* (introduction by F. S. C. Northrop)
❑ Julian Huxley—*Evolutionary Humanism*
❑ Edward Jenner—*Vaccination against Smallpox*
❑ Johannes Kepler—*Epitome of Copernican Astronomy* and *Harmonies of the World*
❑ Charles Mackay—*Extraordinary Popular Delusions and the Madness of Crowds*
❑ Isaac Newton—*The Principia*
❑ Louis Pasteur and Joseph Lister—*Germ Theory and Its Application to Medicine* and *On the Antiseptic Principle of the Practice of Surgery*
❑ Alfred Russel Wallace—*Island Life*

SOCIOLOGY

❑ Emile Durkheim—*Ethics and the Sociology of Morals* (translated with an introduction by Robert T. Hall)